Animal Behaviour: *a concise introduction*

MARK RIDLEY
MA, DPhil,
Astor Junior Research Fellow,
New College, Oxford.
Animal Behaviour Research Group,
Department of Zoology,
University of Oxford.

BLACKWELL SCIENTIFIC PUBLICATIONS
OXFORD LONDON EDINBURGH
BOSTON PALO ALTO MELBOURNE

Editorial offices:
Osney Mead, Oxford, OX2 0EL
8 John Street, London, WC1N 2ES
23 Ainslie Place, Edinburgh, EH3 6AJ
52 Beacon Street, Boston,
Massachusetts 02108, USA
667 Lytton Avenue, Palo Alto,
California 94301, USA
107 Barry Street, Carlton,
Victoria 3053, Australia

First published 1986

Phototypeset by Getset (BTS) Ltd,
Eynsham, Oxford

Printed in Great Britain at the
University Press, Cambridge

DISTRIBUTORS

USA and Canada
Blackwell Scientific Publications Inc
PO Box 50009, Palo Alto
California 94303

Australia
Blackwell Scientific Publications
(Australia) Pty Ltd
107 Barry Street,
Carlton, Victoria 3053

British Library Cataloguing in
Publication Data

Ridley, Mark
Animal behaviour: a concise introduction.
1. Animal behaviour
I. Title
591.51 QL 751

ISBN 0-632-01416-4

Contents

Preface

What follows is an essay in introduction. I have made very few assumptions of my reader, even to the point of briefly explaining how a neuron works and the fundamentals of Mendelian heredity. It has been my aim to introduce the main principles of the modern science of animal behaviour, but I have kept unrelieved discussions of abstract principle to a minimum. I find it easier to pick up principles from examples, and have therefore written the book mainly as a series of exemplary studies of animal behaviour. Although I have not ignored unsettled questions, I have emphasized positive knowledge, in the belief that an introductory work should say something about what is known, in addition to what questions are being asked, and how they can be answered. Because it is a short work, on the 'essentials' of the subject only, I have concentrated on what are, I think, the most important discoveries, to the exclusion of much material dear to my heart. I have preferred to explain relatively few examples at length, filling in a little of the natural history background, rather than covering more examples superficially. This is an introduction, not an encyclopaedia nor a refresher course, and I have not hesitated to reach for classic examples when they can explain a principle most clearly. In short, I have selected my material on the twin grounds of intelligibility and importance. I wish to be understood, and worth being understood.

1 / The biology of behaviour

1.1 Why do animals behave?

All living species of ants live in 'social' groups; these are colonies containing a number of individual ants. Ants characteristically co-operate with other members of their own colony. They may co-operate in catching and transporting their prey or other resources to the nest, they may co-operatively look after the eggs and larval ants in the nest, and they may co-operatively defend the colony from its enemies. Ant colonies, like all living things, have enemies: predators which would feed on the colony, parasites which would exploit it, and other enemies, particularly other colonies of ants, which would take over or steal the resources of the nest. If the colony is to survive, the enemies must be deterred. Most ant colonies possess a special caste of soldiers to do just that. The soldiers of many species have enlarged mouthparts for crushing or snipping, together with a large head to house the muscles that power the mouthparts. Ants use chemical defences too. They may spray poisons, or glues. In the case of the Malaysian seed-gathering ('harvester') ant (closely related to the species named *Camponotus saundersi*), which has been studied by Ulrich and Eleanore Maschwitz, glues are not merely sprayed; when an ant is disturbed it actively explodes, covering any nearby enemies with a fall-out of glue. This species has an enormously enlarged mandibular gland, where the glue is manufactured (Figure 1.1), many times the size of that found in other ant species. A harvester ant in trouble, for instance when it is fighting another ant, explodes itself by contracting the muscles of its abdomen with sufficient strength to split its cuticle, and the sticky contents of the mandibular gland burst out (Figure 1.2) and the glue entangles and immobilizes the victim. The self-sacrifice of the exploded ant should benefit its colony by knocking enemies out of action; but at the cost of its own life.

Why should ants sacrifice themselves in this way? Moreover, this is not the only kind of self-sacrifice to be found among ants. There is also, for instance, reproductive self-sacrifice. Typically only one individual in an ant colony — the queen — breeds; all the other colony members are sterile 'worker' ants, whose business in life is to enable that one reproductive queen to reproduce as much as possible. The queen is an almost static egg machine. She does not feed herself; food is brought by her workers. Nor does she defend herself from enemies; this is also done by her workers. Each egg, after it has been laid by the queen, is carried away by the worker ants, and the egg and young larvae are protected, fed, and reared by the workers. Within the living world

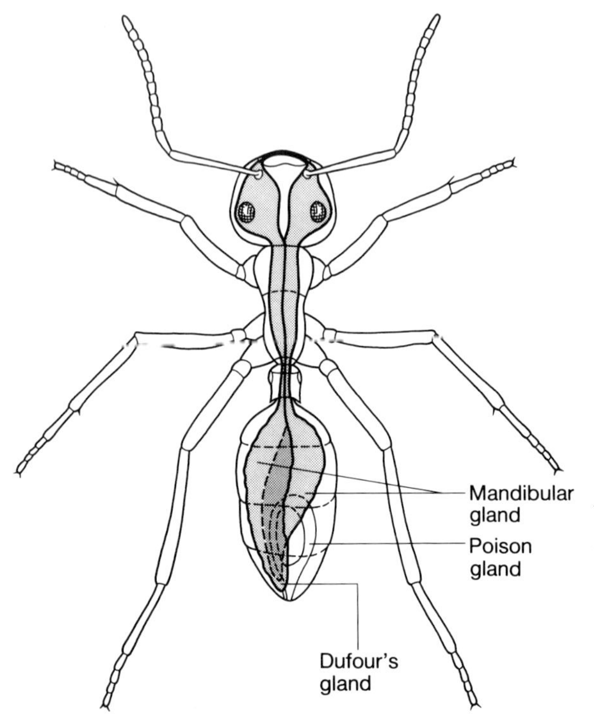

Figure 1.1 The harvester ant *Camponotus* (species near *saundersi*) (left) possesses enormously enlarged mandibular glands. In a more typical ant species, such as the *Iridomyrmex humilis* illustrated in Figure 6.4 (p. 128) the mandibular gland occupies a relatively small part of the head; in this *Camponotus* species the paired gland extends into the abdomen, where it takes up much of the space. (After Maschwitz and Maschwitz, 1974)

as a whole, this arrangement, of many sterile individuals working to increase the reproduction of another member of the species, is exceptional. Species of ants, bees, wasps, and termites have sterile workers, as do a couple of species of aphid. Other examples have been suggested, and no doubt the list will be extended, but in the great majority of species all individuals are capable of reproducing, none are sterile (which is not to say that all individuals of the species *do* reproduce — obviously many die before maturity). Why should ants have sterile castes?

The exact opposite of self-sacrifice for others is selfish exploitation. Blackheaded gulls (*Larus ridibundus*) breed in large and dense groups, also called colonies, but a gull colony, unlike an ant colony, is made up of many familial reproductive units of one male, one female, and their eggs or young. The individuals of a gull colony do not co-operate — quite the reverse. They pursue their own advantage without taking any account of the advantage of other members of the colony. They will, for example, eat each others' offspring (Figure 1.3). When a gull chick hatches from its egg it is, in the words of Richard Dawkins, 'small and defenceless and easy to swallow'. If the chicks

Figure 1.2 When a worker of this *Camponotus* species is disturbed, it contracts its abdomen until it splits; the sticky contents of its mandibular gland then explode out. The ant is here held in a pair of forceps. (Photo: Ulrich Maschwitz)

become separated from their parents, other gulls in the colony may well take advantage of the easy meal, and eat them. Why are some animals as apparently altruistic as the auto-destructing harvester ant, but others as ruthlessly selfish as the cannibalistic blackheaded gull?

If, in a species like the blackheaded gull, a young bird risks death when it goes near adults of its own species (other than its parents), it is clearly important that it should stay close to its own parents. In order to do so, it must be able to distinguish its parents from other objects in its environment, and coordinate its powers of movement with its abilities of recognition in such a way that it does not wander towards moving objects other than its parents. The

Figure 1.3 Colonially nesting gulls are notorious cannibals. Small undefended chicks are taken in large numbers by the adult gulls of the colony. Different adult gulls specialize in hunting different kinds of food, and some individuals specialize in cannibalism.

parent, too, must be able to distinguish its chicks from other chicks, in order to decide correctly which chicks to eat and which to look after. Parents and chicks do have these abilities. The abilities have been studied in some species of gulls and also, among other species, in mallard ducks. The mallard duckling faces a similar worldly problem to a gull chick: its world is made up of one (for the duckling) individual that cares for it, and many other living and non-living objects that are at best indifferent and at worst actively, lethally, hostile. The duckling must distinguish the former from the latter, and stay near the former. Ducklings do in fact stay near their mothers: but how do they manage to do so? If we watch the mallards of a pond in the breeding season we see particular ducklings following particular adults. What process causes the duckling to grow up to follow one individual rather than another? What are the distinctions it is making, and how does it make them?

Consider another example of animal behaviour. A salmon (such as the Atlantic salmon *Salmo salar* which breeds on both sides of the North Atlantic), typically is born (hatches from an egg) in a river tributary. There is no question of parental recognition in salmon because the parents are already dead by the time the egg hatches, but they do have another problem of recognition. After spending a few months in the river, the salmon migrate to the sea where they actively swim around for a year or two, during which time they will swim in many unpredictable directions through thousands of miles. They then finally return to freshwater to breed. Their famous skill — apart from their (in human terms) heroic journey upstream to breed and then inevitably die — is to be able to return to the exact river tributary in which they were born. This fact was first established quantitatively by observations of the Atlantic salmon that breed in the River Tay, in Scotland. Starting in 1903, 550 young salmon smolt were marked with silver wire in their natal river before swimming downstream. The number of marked adult salmon later returning to the Tay and to other nearby rivers was recorded. Of 110 marked salmon, all 110 returned to the Tay: no marked salmon had strayed to other rivers. Since then larger studies have been made. The largest experiment was on the Atlantic salmon of the River Miramuchi, in New Brunswick. Stasko and his colleagues marked 174,509 smolt by fin-clipping; and the 2425 adults later recovered were all taken in the natal stream. As in the River Tay experiment, 100% of the salmon homed perfectly. In other experiments, not all the adults have been recaught in their natal stream; but the stray rates are low, about 2–3% of the total number of adults. The conclusion is that salmon find their way home with high accuracy. A salmon swimming off the Atlantic coast of North America will have a choice of hundreds of freshwater outlets into the sea; and even within its natal river it will have to choose correctly among many branches and sub-branches in order to re-locate its natal tributary. This is an astonishing ability, perhaps because a human would be incapable of it (even if we could swim underwater continually for a year or two). How do salmon find their way home?

Such questions as why ants sacrifice themselves for their colony, whereas blackheaded gulls would sacrifice their colonies for themselves; or, why does a mallard duckling follow one particular adult mallard rather than another; or how does a salmon find its way home, are all examples of the questions we shall be seeking to answer in this book. They are also the kinds of question that make up the science of animal behaviour. Ethologists, behavioural ecologists or sociobiologists (there is no universally accepted term for the whole subject) aim to discover how animals do behave, and then ask, and answer, questions about why the animals behave the way they do.

The general question of animal behaviour — 'Why do animals behave?' — looks like a single kind of question; but when it is replaced by a series of concrete questions, such as those we have just posed, several different kinds of question are revealed. If we ask why a harvester ant explodes, we might expect an answer in terms of the neuronal impulses to its abdominal musculature which cause its cuticle to split and release the sticky contents of its mandibular gland. We are then treating the ant as a machine, and asking what mechanisms it uses to produce its behavioural output. But another kind of answer might be returned to the same question. We might reply that the harvester ant explodes in order to defend its colony. This answer is of an independent kind, because its truth does not depend on any particular idea about the exact neuromuscular mechanism used to detonate the explosion; and vice versa the analysis of the mechanism is independent of what function the behaviour performs. We can work out the mechanism whether or not we have identified the function correctly — whether the contents of the mandibular gland are a sticky defensive chemical (which in fact they are) or something else, such as food (which in fact they are not).

There are more than these two types of question and answer. Consider next our questions about ducklings — how does a particular duckling come to follow one adult rather than another? The same two kinds of answer are again possible (in terms of bodily mechanisms and functional consequence), but the natural answer is now of developmental origin. Ducklings, as it happens, learn to follow whatever object they see near them during a 'sensitive' phase after they hatch. That object is normally their mother, and different ducklings follow different adults because they see different adults during that sensitive phase.

A fourth (and final) kind of answer to the general question of why animals behave is analogous to the developmental answer, but over a much longer time scale. We can explain differences in how animals behave, as we did for ducklings, in terms of their different life-time experiences: we can do likewise for their different, inherited, evolutionary ancestries. Thus, if we wish to explain why, say, the species of fish called three-spined sticklebacks (*Gasterosteus aculeatus*) court with their particular behavioural sequence (p. 167), but mallard ducks court with another quite different kind of behaviour (pp. 16–17), we might say that sticklebacks are descended from ancestors that performed stickleback-like courtship dances, mallards from ancestors that performed mallard-like courtship, and that each of the modern forms has inherited the habits of their different ancestors.

Modern students of animal behaviour, following an analysis of Niko Tinbergen (although the relevant distinctions had been made by Aristotle), thus customarily distinguish four kinds of question and answer, rather than a simple general one, to the question of why animals behave. The four are separate in the sense that they are independent of, but compatible with, each other: independent because any given answer to one question does not imply any particular answer to any of the other three; compatible because all four can in principle be asked of any unit of behaviour. The form of a question may invite one of the four kinds of answer — as we saw when asking why different ducklings follow different adult ducks, but the behaviour can still be analysed in the other three ways. We can still ask what mechanisms the duckling uses to distinguish adults, to establish its own preference for one particular adult, and to translate the perceptual distinction and learned preference into action; we can still ask about the ancestral history of filial responses in ducks, and birds in general, and their advantages or purposes in the lives of those birds.

It is worthwhile to distinguish the different kinds of question, for it is a fact of history that students of animal behaviour have frequently argued at cross-purposes by confusing them. The first two, concerning mechanisms and purposes, have been confused particularly often. However, now the distinction has been made it causes little difficulty, for it is, in truth, very simple. I therefore have established it at the outset less because of any pressing urgency than because it provides a convenient way of thinking about animal behaviour as a whole and in its parts. Before we start to consider the four questions in detail, we should discuss one other fundamental point. What is the nature of behaviour? How can it be defined and recognized? For if different observers do not agree on what behaviour is, the scientific study of behaviour will be impossible.

1.2 What is behaviour?

The simplest definition of behaviour is movement, whether it is the movement of legs in walking, wings in flying, or heads in feeding. But some actions of animals, such as the honking of peacocks, which we should wish to count as behaviour, are not movements of the whole animal in the ordinary sense. The honking sound is produced as air is forced, by the contraction of muscles, out of the peacock's lungs, which causes a region of the throat to vibrate. There is movement here, of the pulmonary musculature, just as there is muscular movement when an animal feeds or walks: in a more accurate sense, therefore, animal behaviour consists of a series of muscular contractions.

Naturalists had recorded incidental observations of behaviour for many centuries, but no real attempt at the scientific study of behaviour was made earlier than about a century ago. A crucial insight of the earliest workers — Charles Darwin, Oskar Heinroth, Konrad Lorenz — was that behaviour is orderly enough to allow that necessary criterion of all science – repeated observation. Behaviour, or muscular contractions (they noticed), comes in orderly sequences, recognizable patterns of behaviour which can be called behavioural 'units'. The same animal will produce the same pattern of movements again and again; different members of the same species will also behave in recognizably similar ways. Behaviour can only be studied because of this fact. It makes it possible for an observer to check his own evidence, and for different observers to check each others' evidence. Without recognized units of behaviour, anecdotes might accumulate, but each would be closed to criticism, and rigorous testing of theories would be impossible.

Is it correct to assert that behaviour is so regular? Let us consider an example. We can illustrate the principle by that classic example of a be-

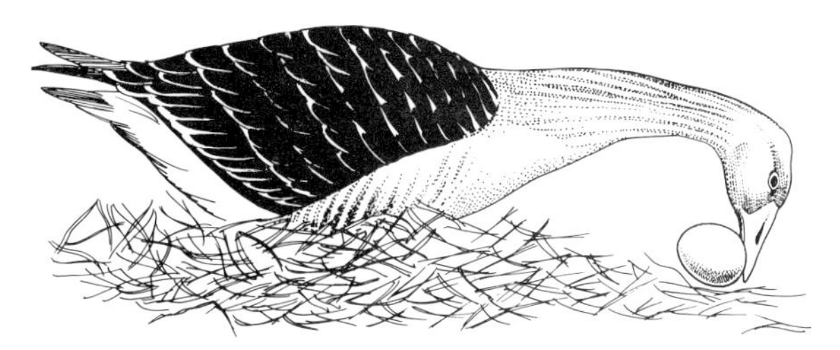

Figure 1.4 Egg retrieval by greylag goose. The gull in Figure 2.9 below (page 41) is carrying out a similar task, in an experiment. (After Lorenz and Tinbergen)

havioural unit, the 'egg retrieval response' of the greylag goose. The greylag goose, which was Konrad Lorenz's favourite study animal, breeds in monogamous pairs. It nests on the ground, the nest being little more than an area of grass shaped into a bowl with the edge built up, though not enough to prevent an egg from occasionally rolling out. This is the occasion for the egg-retrieval response (Figure 1.4). When a goose sees an egg just outside its nest, it enacts the following sequence of muscular movements. Standing in the nest, it first extends its neck outwards until its head is above the egg. It then puts the underside of its bill against the further side of the egg, and starts to roll it back. While rolling the egg, the goose moves its bill from side to side, to prevent the egg from slipping away to the side. The behaviour is not always effective, the egg may slip away. When it does, the goose does not immediately stop moving its bill backwards and re-establish contact with the egg. Instead it moves its bill all the way back to the nest and only then, when it again sees an egg (in fact the same one) outside the nest, does it place its bill against the egg, and try again. In other words, once started, the behavioural unit is continued until it is finished. Moreover, when Lorenz removed the egg from a goose while she was in the middle of rolling it back, the goose still continued and completed the sequence of movements. The two observations prove that sensory feedback, of the feel of the egg against the bill, is not needed to stimulate the continuing movement of the neck muscles.

Behaviour patterns can be recognized as units if they are performed often enough, and in similar enough form. Of course, the egg-retrieval response of different geese, and of the same goose on different occasions, will not be exactly identical. But identical repetition is not necessary to define a unit of behaviour. The behaviour pattern on different occasions only needs to be sufficiently similar to be recognizable, and then the behaviour units can be defined statistically.

Behaviour can be described inconsequentially, as a series of movements, as we have just done for the egg-retrieval response of the greylag goose, but it can also be described in terms of its consequences. 'Retrieve egg', for instance, is a consequential description; it does not mention the exact movements used, but does specify what results from them. For the general point being made here, it does not matter much which method we use to describe behavioural units; all that matters is that animals perform behaviour patterns which can be recognized by different observers. Then scientific study becomes possible.

The point can be made another way. Darwin, Heinroth, Lorenz, and other early students of behaviour, would have been educated to think biologically about the anatomical parts of animals, rather than their behaviour. They would have learned how to study parts such as limbs, urino-genital systems and circulatory systems. When they came to think about behaviour they naturally conceived and described behaviour in the form of units, rather analogous to the limbs, kidneys and hearts that can be seen in the anatomy of an animal. Feeding, or courtship, in animals can be studied in the same way as their anatomy and physiology. We could ask the same kind of questions about them. Indeed, Heinroth and then Lorenz both started their work on animal behaviour by applying biological methods, such as tracing the course of development of units of behaviour in the life of an individual, and looking at different species and trying to see the equivalent units of behaviour in all of them. For it is not only the greylag goose that uses that recognizable egg-retrieval response. A similar sequence of muscular contractions is used by all other ground-nesting bird species to retrieve their eggs. For example, other species of geese and all species of gulls show an egg-retrieval response which is recognizably the same unit of behaviour. But modern ethology is different from Konrad Lorenz's early work, and it has moved on to more fascinating units of behaviour than the egg-retrieval response. However, this response illustrates the basis of the possibility of a scientific approach: recognizable sequences of muscular contractions and recognizable units of behaviour.

We have now said enough about the nature of the science and can move on to consider its content. Tinbergen's distinction of the four 'whys', the four different kinds of question that people ask about animal behaviour, is a useful structure for the material. Thus, Chapter 2 will discuss the question of mechanism, and Chapter 3 that of development. But evolution and natural selection are of such all-pervasive importance that we must consider them, in relation to animal behaviour, immediately. Then, equipped with these fundamental concepts, we can move on to apply them to the behaviour of animals, as they maintain themselves against their environments, and against the society of members of their own species.

1.3 Evolution

Living things, as simple observation will reveal, come in distinguishable, recognizable forms such as robins, thrushes and blackbirds. These are what biologists call species. A species is a group of organisms that can breed with one another to produce another member of the same species; robins breed with robins, not blackbirds. All the offspring of robins are obviously robins; you could follow all the generations of robins produced during a human lifetime, and all of them would look like robins. This constancy of species is probably the reason why the idea of evolution only took root relatively recently. It is so easy to extrapolate from one's limited experiences, and conclude that robins always have been descended from ancestral robins back through eternity. One would thus arrive at the theory — still believed by some — that species are fixed, and each species has a separate ancestry.

But although most species are constant in form over the temporal and spatial scale of normal human experience, as the range of evidence is expanded, the constancy breaks down. It is difficult to expand the time scale of evidence, because the only obvious means is by examining fossils, and the fossil record is too poor to allow us to trace the ancestry, through a continuous series of fossils, of more than a few exceptional species. But geographical travel, to expand the scale of evidence in space, does break down the impression of the constancy of species. At any one place, species do appear as discrete groups of organisms; but if one traces a species across the world it will usually change considerably. The house sparrows of North America, for instance, differ from those of Europe; and the herring gull and black-backed gull, which behave as two perfectly ordinary species in Britain, are connected by a continuous ring of intermediate forms around the North Pole. Geographical variation first led Darwin to doubt the constancy of species, particularly after he had returned from the circumglobal voyage of the *Beagle*. He had mixed his collections of finches from the various islands of the Galapagos archipelago, because he thought they were all one form; but he soon realized that different species of finch inhabited the different islands. It was a striking discovery. How easy it now was to imagine that the different species of Galapagos finch had evolved from a common ancestor, rather than being separately created on their respective islands! For why, on the Galapagos, should the warbler-like birds and the woodpecker-like birds, as well as the finch-like birds, all have been created as finches, when in the rest of the world ordinary warblers and woodpeckers arose? (Figure 1.5.)

For this and other reasons, almost all biologists think that species are not fixed. They think instead that species change slowly, over long periods of time. Therefore, if we traced back the ancestors of robins, we should come to forms that, although recognizably birds, were not robins; and as we traced the lineage to earlier and earlier ancestors we should come to forms that were more like amphibians than birds, and eventually to animals without backbones (invertebrates). Their very early ancestors were simple cells, which floated in the sea some 3500 million years ago. These cells, or perhaps some earlier form of life, were the common ancestor of all the species of animals and plants now alive: at least, there is evidence that suggests that all modern forms share a single common ancestor. According to the theory of evolution, species have taken on their present appearance as they have changed from their ancestors.

The behaviour of animals has presumably changed in evolution, just as their anatomical appearance has. But for the hard parts of organisms we

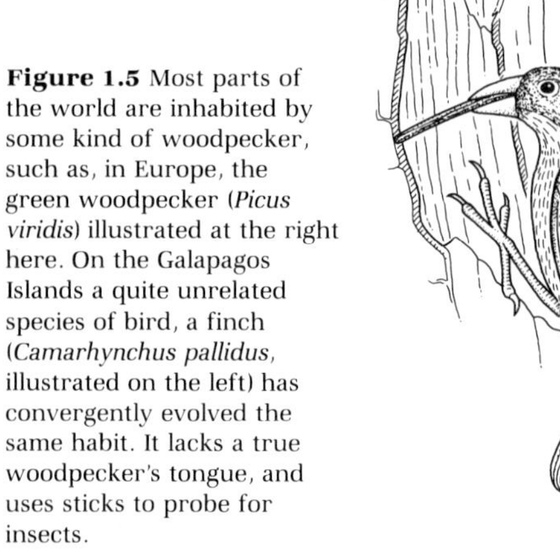

Figure 1.5 Most parts of the world are inhabited by some kind of woodpecker, such as, in Europe, the green woodpecker (*Picus viridis*) illustrated at the right here. On the Galapagos Islands a quite unrelated species of bird, a finch (*Camarhynchus pallidus*, illustrated on the left) has convergently evolved the same habit. It lacks a true woodpecker's tongue, and uses sticks to probe for insects.

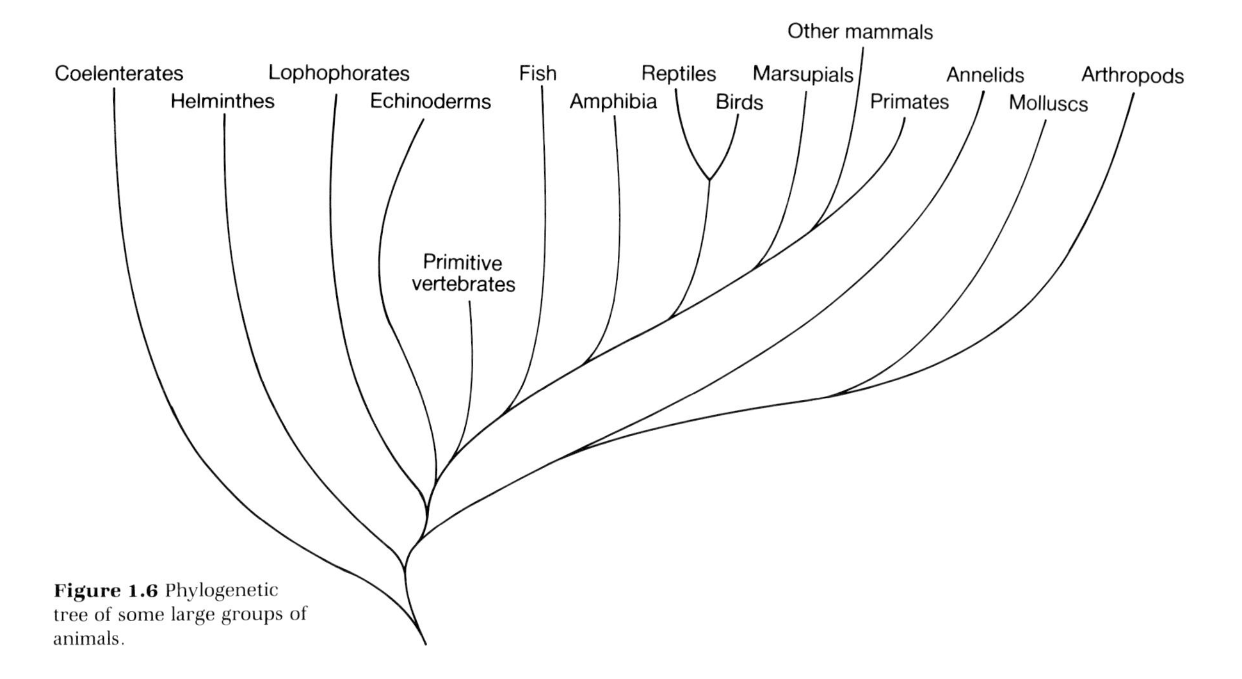

Figure 1.6 Phylogenetic tree of some large groups of animals.

possess some evidence from fossils as to what they were like in the past which is not available for behaviour. Fossilized animals are necessarily dead, they do not behave, and although it is often possible to infer something about how a fossilized animal would have behaved, the inference is uncertain and superficial.

This does not mean we know nothing about the evolution of behaviour. It merely means that our knowledge comes from another kind of evidence. In fact, the evolution of behaviour is studied by comparing the behaviour of different living species. Let us consider how this knowledge can be acquired. The first requirement is a phylogenetic tree, which shows the ancestral relations of modern forms (Figure 1.6). We still know little about the phylogenetic relations of many animal groups, particularly of invertebrates; but the general relations, and even the finer connections of some well-studied groups like birds and mammals, are reasonably well known. Phylogenetic trees like Figure 1.6 are mainly inferred from morphological and molecular evidence. Consider, for example, three of the groups of Figure 1.6: the birds, amphibians, and echinoderms. Two of them (birds and amphibians) resemble each other more closely in their molecular structure and their morphology (for example they possess a hollow dorsal nerve chord, a backbone, and segmented muscles), than either does with the third group (the echinoderms). We infer that the birds and the amphibians share a more recent common ancestor with each other than either does with the echinoderms, and the birds and the amphibians are therefore put together in the phylogeny. Actually, this statement grossly simplifies the modern methods of phylogenetic inference, which do not work simply from the similarity of groups to each other; but the important point here is that because the methods of phylogenetic inference make little or no use of behavioural evidence, it is possible to use established phylogenetic trees to study the evolution of behaviour without falling into a circular argument.

Given a knowledge of phylogeny, we can immediately infer whether a behaviour pattern that is found in more than one group of animals has evolved independently in each of them (which is called convergent evolution) or evolved only once, and is now shared in different groups by descent from a common ancestor. The rule is, that if the different groups in question lie far apart in the phylogenetic tree, and are separated by many groups with different habits, the behaviour pattern is probably convergent. Take the habit of monogamy, and biparental care (both male and female caring for the young). This is found in many species of birds, and it may have evolved only once; but it is also found in (among other odd groups) some species of cichlid fish. Because most of the groups of animals between cichlids and birds on the phylogenetic tree (i.e. most other fish, amphibians and some reptiles) lack monogamy and biparental care, it probably evolved independently. It is convergent in the two.

More striking examples of convergence can be found in the social insects. Social life itself, with a sterile worker caste, has evolved independently in the termites and the social hymenopteran groups (the ants, bees and wasps, among which the social habit is thought to have evolved independently at least eleven times). Many strikingly similar social habits have evolved in ants and termites. For instance, species of both groups have independently evolved soldier castes, and in both cases the soldiers have evolved enlarged heads and mouthparts (Figure 1.7). There is even a termite species that has evolved the same habit as the exploding harvester ant. E. O. Wilson describes the soldiers of the termite species *Globitermes sulphureus* as

'quite literally walking chemical bombs. Their reservoirs fill the anterior half

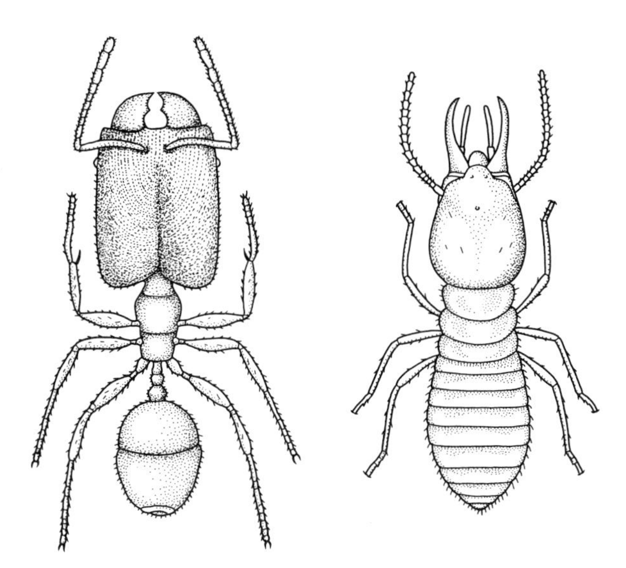

Figure 1.7 The 'soldier' castes of many species of ants and termites are an example of evolutionary convergence; they have evolved independently in the two groups. The soldiers in both the species illustrated here have enlarged mouthparts for gripping and crushing. The ant species is *Pheidole tepicana* (left), the termite species *Prorhinotermes simplex* (right). Soldiers of other termite species possess a 'nasus', a chemical spray gun, on their heads. The form of the soldiers' adaptations vary among species; but in all cases they are adapted to defend their nest mates from live enemies. (After Wheeler and Banks & Snyder)

of the abdomen. When attacking, they eject a large amount of yellow liquid through their mouths, which congeals in the air and often fatally entangles both the termites and their victims. The spray is evidently powered by contractions of the abdominal wall. Occasionally these contractions become so violent that the wall bursts, shedding defensive fluid in all directions.'

The habits of the two species are similar (although there are some minor differences), but have certainly evolved separately. The reason for supposing that the habit is convergent is as follows. If it were not convergent the common ancestor of termites and ants would have had the explosive defensive habit; so too then would the continuous series of thousands of species (of which there are hardly any remains) connecting the modern ants with the modern termites. Every species of insect phylogenetically intermediate between ants and termites would then have been descended from a common ancestor with the explosive habit, but in all those species the habit must have been lost — no other living insects except *Camponotus saundersi* and *Globitermes sulphureus* are known to explode in this way. We should thus have to infer a large number — perhaps many thousands — of independent losses of the habit. Whereas, if the habit is convergent in the two species, we do not need to invoke any (now unknown) intermediate ancestors with the habit, nor all those thousands of evolutionary losses. It is therefore simpler to suggest convergence.

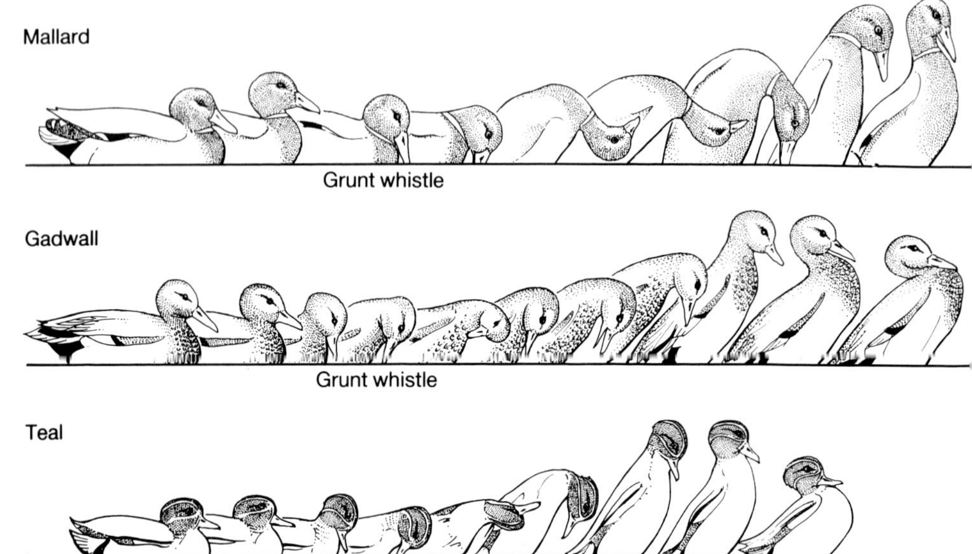

Figure 1.8 The courtships of male mallards (top), gadwall (middle), and teal (bottom) are made up of sequences of distinct displays. Konrad Lorenz observed the courtship of different duck species and compared the displays: the comparison among species reveals the nature, if not the direction, of their evolutionary modification. (From Lorenz *The evolution of behavior* Copyright © 1958 by Scientific American, Inc. All rights reserved.)

So much, for now, for convergence. If we are studying a habit shared by a group of closely related species we can carry out another kind of analysis. The courtships of ducks are a good example. They were first studied systematically by Heinroth, and later by Lorenz. Lorenz observed the pattern of courtship, and other kinds of social behaviour in twenty species of ducks and geese. He divided the behavioural repertory of each species into behavioural units. Some of the habits are only found in a few species, but others, such as 'monosyllabic lost piping' (a distress call of the chicks), were found in all species. Figure 1.8 illustrates how the same unit may be recognizable in different species. Once this has been recognized, two kinds of inference become possible. One is to infer the evolutionary changes undergone by the behavioural unit. We can see, for instance, how the 'head-flick' display differs among three species in the figure; to take one feature, the gadwall appears to perform this display nearer the water than does the mallard. Any behavioural unit, although it may be recognizable in more than one species, will not be exactly the same in all cases. Just as anatomical units, such as the limb bones, can be traced from the fins of fish to their very different form in the legs of birds and reptiles, so can the evolutionary changes of behaviour be traced by comparison among species. It is sometimes difficult, or even impossible, to discover the direction of the changes through

evolutionary time; but inferences can be made by the same techniques as are used with anatomical evolution. In either case, the nature, if not the direction, of change can be worked out.

The second kind of inference puts into reverse an earlier statement. I said that our phylogenetic knowledge is derived from anatomical and molecular evidence; this is true, but there is no reason why behavioural evidence should not be used in the same way. If the phylogenetic relations of species are suggested by the similarity of their morphological appearances or the sequences of their proteins, they can also be suggested by the similarity of their behaviour. Lorenz used this fact in his work on the behaviour of ducks and geese. He had recorded a total of forty-eight different behavioural units in twenty species. He could then group the species according to how similar they were with respect to these units. In Figure 1.9 the vertical lines represent the twenty species, which have been arranged according to shared behaviour patterns (indicated by the horizontal lines). If we thought that behavioural similarity between two species implied a recently shared evolutionary ancestry, then we could infer from Figure 1.9 that, for instance, the pintail and the mallard share a more recent common ancestor than either do with the shelduck. The grouping of the species according to their behaviour is similar to, but not exactly identical with, their grouping by their anatomical

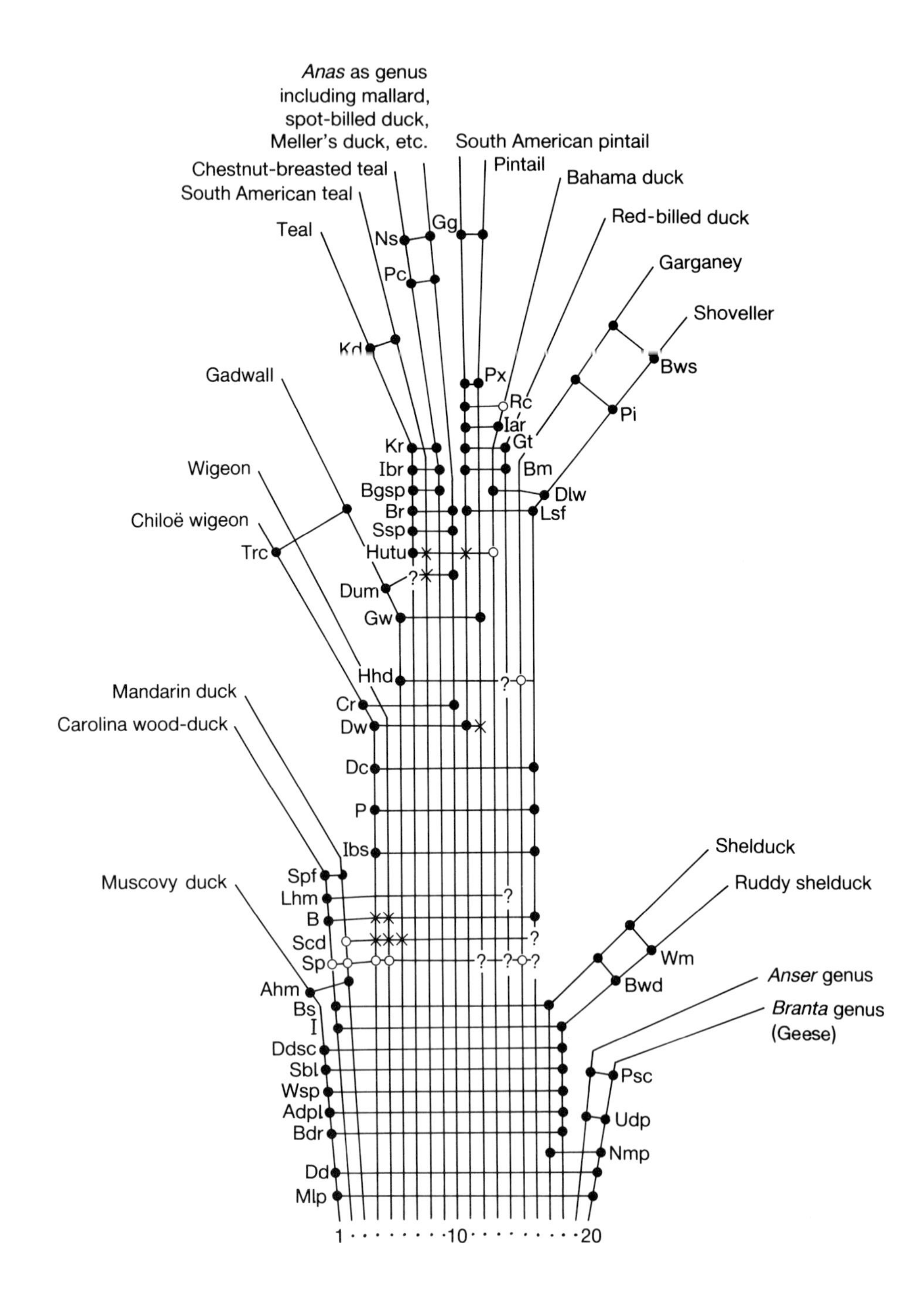

Anas as genus including mallard, spot-billed duck, Meller's duck, etc.
South American pintail
Pintail
Chestnut-breasted teal
South American teal
Bahama duck
Red-billed duck
Teal
Garganey
Shoveller
Gadwall
Wigeon
Chiloë wigeon
Mandarin duck
Carolina wood-duck
Muscovy duck
Shelduck
Ruddy shelduck
Anser genus
Branta genus (Geese)
Ns
Gg
Pc
Kd
Px
Rc
Iar
Gt
Bws
Pi
Kr
Ibr
Bgsp
Br
Ssp
Hutu
Bm
Dlw
Lsf
Trc
Dum
Gw
Hhd
Cr
Dw
Dc
P
Ibs
Spf
Lhm
B
Scd
Sp
Ahm
Bs
I
Ddsc
Sbl
Wsp
Adpl
Bdr
Dd
Mlp
Wm
Bwd
Psc
Udp
Nmp
1 · · · · · · · · 10 · · · · · · · · · 20

Figure 1.9 Shared behavioural patterns can be used to classify species in the same way as morphological traits. The figure shows Lorenz's classification of twenty species of ducks by their similarities with respect to forty-eight behavioural traits. The names of the behavioural traits are as follows:

Mlp monosyllabic 'lost-piping'
Dd display drinking
Bdr bony drum on the drake's trachea
Adpl Anatine duckling plumage
Wsp wing speculum
Sbl sieve bill with horny lamellae
Ddsc disyllabic duckling social contact call
I incitement by the female
Bs body-shaking as a courtship or demonstrative gesture
Ahm aiming head-movements as a mating prelude
Sp sham-preening of the drake, performed behind the wings
Scd social courtship of the drakes
B 'burping'
Lhm lateral head movement of the inciting female
Spf specific feather specializations serving sham-preening
Ibs introductory body-shaking
P pumping as prelude to mating
Dc decrescendo call of the female
Br bridling
Cr chin-raising
Hhd hind-head display of the drake
Gw grunt-whistle
Dum down-up movement
Hutu head-up-tail-up
Ssp speculum same in both sexes
Wm black-and-white and red-brown wing markings of Casarcinae
Bgsp black-gold-green teal speculum
Trc chin-raising reminiscent of the triumph ceremony
Ibr isolated bridling not coupled to head-up-tail-up
Kr 'krick'-whistle
Kd 'koo-dick' of the true teals
Pc post-copulatory play with bridling and nod-swimming
Ns nod-swimming by the female
Gg *geeeeegeeeee*-call of the true pintail drakes
Px Pintail-like extension of the median tail-feathers
Rc R-calls of the female in incitement and as social contact call
Iar incitement with anterior of body raised
Gt graduated tail
Bm bill markings with spot and light-coloured sides
Dlw drake lacks whistle
Lsf lancet-shaped shoulder-feathers
Bws blue wing secondaries
Pi pumping as incitement
Dw drake whistle
Bwd black-and-white duckling plumage
Psc polysyllabic gosling social contact call of Anserinae
Udp uniform duckling plumage
Nmp neck-dipping as mating prelude

(after Lorenz).

similarity. Lorenz's grouping of the ducks and geese therefore resembles the more traditional classifications.

We need not enter into details, for this kind of study has not turned out to be of much importance in the science of animal behaviour. Many of Lorenz's methods of studying animal behaviour have been followed by later workers, but his use of behaviour to discover phylogenetic relations is not generally one of them. The study of behaviour has tended to exploit, rather than increase, our knowledge of phylogeny. However, even if Lorenz's phylogenetic work has not been influential, it still matters. It illustrates how the evolution of behaviour may be studied; it also shows again how his early ideas on how behaviour should be studied were influenced by the work of his contemporaries on the hard parts of animals.

1.4 Natural selection and adaptation

After we have accepted that behaviour has evolved — that it has changed over evolutionary time — the next question is why it has evolved through the course that it has. The question is a special case of a more general question, of why evolution occurs at all. The generally accepted answer, for the majority of if not all evolutionary change, is the principle discovered by Darwin and called by him natural selection. If we first establish what natural selection is, it will become fairly obvious why it drives evolution. And we can then consider the importance of the principle for animal behaviour.

Natural selection is easily understood. Its operation follows logically from a series of elementary propositions. The first is that species vary. If you were to measure some trait in many members of a species, for instance the body weight of male Hercules beetles, it would almost certainly show variation — this is true of nearly all traits in all species, everywhere. The second proposition is that the variation is often, to some extent, inherited. Larger male Hercules beetles may tend to produce larger than average offspring (although, so far as I know, this particular trait has never been studied). The third proposition is that the members of all species produce very many more offspring than can ever survive. Female cod produce millions of eggs, but even the larger mammals, which produce only one offspring every year or two, still produce many more offspring than could survive. Darwin calculated, for instance, that a single pair of elephants could have 19 million descendants alive about 750 years after their birth. Clearly, whether in cod or elephant, most of the offspring must die. If any inherited variant of the species is, in the slightest degree, more likely to survive to reproduce, then evolution by natural selection must operate. That variant will survive better, leave (on average) more offspring (which will resemble their parents) and therefore increase in frequency in the population. The next generation will contain more of it than the previous generation — the population will have slightly altered and evolution will have taken place. If the conditions are maintained, the variant that survives better may continue to increase in frequency until it makes up the entire population.

Natural selection has been observed in operation in many cases, of which by far the clearest example is the research of H. B. D. Kettlewell on the peppered moth *Biston betularia* in Britain. In this famous study there were two main variant types of the moth, a lighter peppered coloured type and a darker melanic type. The difference is inherited; that is, the offspring of the melanic type are more likely to be melanic than are the offspring of the peppered type. The moths are eaten by birds which hunt their prey by sight; each type survives better (because it is less vulnerable to predation) against a different background. The peppered type survives better on lichen-covered trees, and the melanic type on the dark, lichen-less trees of industrially polluted areas (Figure 1.10). When the proportions of the two kinds of tree alter, natural selection causes a change in the proportions of the two types of moth. When, in particular, the proportion of dark trees increased in Britain after the early nineteenth century industrial revolution, the melanic type of moth increased from a rare minority to be the majority form within about 50 years.

There are two conditions for natural selection to drive evolutionary change: the environment itself must change to alter the advantages of the different types; and the difference must be inherited. If the environment does not change, natural selection will probably maintain the species in constant form. Only if the environment changes can natural selection cause evolutionary change; but even then it can only do so if some kinds of individuals of the species survive to reproduce better than others, and if the ability that enables them to survive better is inherited. If the difference between types is not inherited, no evolution will take place even if one type does survive better than another. It could be, for instance, that the bigger members of a species survived better than the smaller ones; but size might not be inherited, that is, bigger than average parents would not leave bigger than

Figure 1.10 H. B. D. Kettlewell (*below*) placed samples of the two types of peppered moth on trees in polluted and unpolluted areas of England. By observing from a nearby hide, he measured the rates at which birds took the two types in different places. Birds, such as robins and redstarts (*above*), took the more conspicuous types at higher rates. Tinbergen collaborated in this work in the 1950s. (Photo: Niko Tinbergen, and predatory birds redrawn from photos by Niko Tinbergen)

average offspring. (Differences in size might be controlled by noninherited accidents, such as how much food an individual happened to find.) In this case, the average size of the species would stay constant even though bigger animals were surviving better. Only if the size of an individual became an inherited property could evolutionary change towards larger size take place. If any trait is to evolve under natural selection, differences in the trait *must* be inherited.

Natural selection is the reason why species evolve, but its importance for the student of animal behaviour lies less in its explanation of evolution than of adaptation. Adaptation, in biology, refers to the fit between an organism and its environment, to the adjustment of the parts of the organism to the needs of its way of life. The behaviour of animals is adapted just as much as their anatomy and physiology. But it is not always obvious how particular habits are adapted to the way of life of the organism. Indeed, many of the most interesting studies of animal behaviour have sought to uncover the advantage of particular habits — be they self-explosion in harvester ants, or extravagant courtship displays in birds of paradise. As an explanation of adaptation, natural selection is the key to one of our four questions about why animals behave.

Not every conceivable kind of advantageous trait can evolve under the power of natural selection. Natural selection works by differential reproduction. To a first approximation, it can only favour traits that increase the number of offspring left by the organism. When we ask why an animal performs a particular behaviour pattern, if we are to return an accurate answer it is necessary to translate the question into 'how does that behaviour enable that kind of organism to produce more offspring?' or 'why would an animal that performed a different behaviour pattern leave less offspring?' If a postulated advantage will not translate into more offspring it cannot be the true explanation of the behaviour. This kind of inquiry is usually a matter of finding out to what particular need or property of the species' life style and environment a behaviour pattern is adapted. In the case of the peppered moth, which we have just discussed, the answer is obvious. The adaptation is the colour pattern, and the properties of the environment to which it is adapted are the treetrunks it rests on and the birds that eat the moths. The reason why a moth of some other colour pattern would leave less offspring than the camouflaged type is, quite simply, that it would be more likely to be eaten before it reproduced. In some of the behavioural adaptations we shall come to, particularly those concerning social behaviour, the advantages are less obvious. But it is necessary to establish the principle by an unambiguous example to begin with.

1.5 Summary

The science of behaviour is concerned with questions of why animals behave; these are a set of different kinds of such questions rather than a single one. Four main kinds of question and answer can be distinguished:
By what mechanism is the behaviour produced?
What use is it to the animal?
How did it develop?
What is its evolutionary history?
The questions are compatible; they can all be asked of any particular behaviour pattern. But they are independent; any particular answer to one of them implies little about the answers to the others. The scientific study of behaviour is made possible by the fact that the behaviour of animals is performed in repeatable, publicly recognizable units.

All modern species have descended from a common ancestral species. During the evolutionary changes of species from their ancestral forms, their behaviour must have changed as well. Evolutionary changes in behaviour are not studied directly, in sequences of fossils, but by the comparison of modern forms, using our knowledge of phylogeny. The course of behavioural evolution can be inferred by comparing the behaviour patterns of species known to be descended from a recent common ancestor. Similar behaviour in different groups of species is inferred to be convergent if the groups are phylogenetically distant, and separated by species showing different behaviour. Evolution has taken place because of the process of natural selection. Natural selection favours those types of animals that leave more offspring than the average for their species. It causes animals to acquire, over many generations, any trait that increases their reproductive powers. The reason why the behaviour of animals is adapted to their environments, therefore, is that it has been established by natural selection, because it enabled relatively more reproduction than less well adapted habits. To understand the behaviour of animals we need to find out how it enables them to reproduce more than they otherwise would.

1.6 Further reading

Tinbergen (1963) originally distinguished the four kinds of questions about animal behaviour; later discussions have not essentially modified his analysis. Barlow (1977) and M. S. Dawkins (1983) discuss how behaviour is organized into units. Darwin's *'Origin of Species'* remains a good summary of the case for evolution: I recommend the first edition (1859), in which the especially relevant chapters are 9–13. I have given a shorter modern summary (Ridley, 1985). Wilson (1971) is a magnificent account of the social insects, which includes a discussion of the soldier caste, and is the source of my quotation. Lorenz (1958) describes his work on the behaviour of ducks. There are many excellent modern accounts of adaptation and natural selection; I recommend, in order of increasing sophistication, R. Dawkins (1976), Maynard Smith (1975) and then Williams (1966).

2 / The machinery of behaviour

In this chapter we shall be treating animals as machines, and seeking to understand how they work. We shall not be completely successful. The behaviour of animals is flexible and diverse. During a typical day an individual animal may perform, in a particular order, dozens of different behaviour patterns, or units. A male bird in the spring, for example, may sing at dawn, and then feed (using many different foraging techniques), hide from predators, defend its territory, and court a female, during one day. Ethologists have not progressed far in describing the mechanisms that control the complete behavioural repertory of an animal through its life — that is, the way it decides what behaviour patterns to perform in a given set of circumstances, how to trade off its motives to perform one activity rather than another, and by what physiological mechanisms these decisions are made.

If ethologists have not progressed far in solving the total problem, they have made many important and curious discoveries about the mechanisms controlling some individual behaviour patterns. This is particularly true of animal senses, as I shall illustrate by a section on echolocation in bats. The general, elemental mechanisms controlling behaviour are known, even if how they are combined to produce the full behavioural output of an animal is not.

Muscular contractions, as we have seen, are an important mechanical source of many behaviour patterns. Animals control their muscles by their nervous systems, and we shall consider how the nervous system works. The nervous system also carries information from the sense organs and, by mechanisms of excitation and inhibition among its different parts, integrates the animal's senses with its internal tendencies and preferences. When it is possible to explain a behaviour pattern in nervous terms, I shall do so. But the study of animal nervous systems has not in all cases turned out to be the easiest route to understanding how they control their behaviour, and we shall consider some examples of how animal senses have been successfully studied at a higher level than that of sensory neurons, and their behavioural choices at a higher level than the influence of neurons on one another.

2.1 Complex results can be produced by simple mechanisms: the spider's web

A human, faced with the task of building some such geometrically complex structure as a spider's web (Figure 2.1) would do so using a blueprint, or plan, of the web. They would then control their behaviour by reference to their

Figure 2.1 Orb webs are built by many species of spiders, such as (left) this British adult female of the common garden spider *Araneus diadematus* (the spider is in the centre, at the hub of the web), and (right) an immature Floridan *Argiope aurantia*. *Argiope* characteristically spin a 'stabilimentum' using special silk at the hub of their webs. The stabilimentum is conspicuous, and has been argued by Tom Eisner to advertise the web to birds, to stop them flying through it unaware; or the stabilimentum may act to disguise the spider, as it does in this photograph. (Photos: Fritz Vollrath)

concept of the goal to be reached. Perhaps the araneid spiders, such as the common garden spider *Araneus diadematus*, which do build orb webs, have a concept of the web; we cannot tell. However, observations of spiders in action suggest that they follow a series of rules, which in themselves would be sufficient to lead the spider to build a web even if it had no idea of what the end point should be. Not all the flexible details of the spider's building program have been worked out, but the main rules were discovered simply by watching them build. The observations are easy to make. Orb webs are

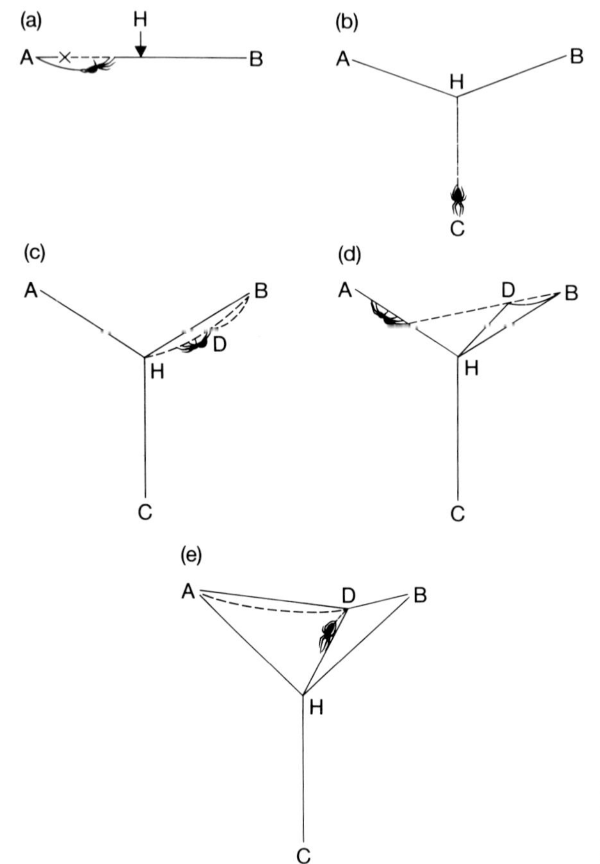

Figure 2.2 The stages in the construction of an orb web. See text for explanation. (After Peters)

spun by common spiders all over the world; and the whole operation takes less than half an hour.

The end product, the orb web, is a regular structure made up of frame, radial spokes, and catching spiral (Figure 2.1). The stages of its construction were first formalized by Hans Peters in 1939, and they are as follows. The spider starts with the frame and spokes. To begin with, she spins a thread with one unattached end and allows it to be blown by the breeze. The other end is attached to the spider, by the spinnerets of the abdomen, which are the source of the thread. The loose end will soon become entangled in some object, such as a nearby twig. The spider then bites through her end of the thread (at A in Figure 2.2a) and, having attached a new thread at her point of departure, walks off down the wind-cast thread, spinning another thread as she goes. When she arrives at the other end (B of Figure 2.2a) she attaches the new thread. She then turns, walks some distance up the thread (to the point H on Figure 2.2b), and attaches a new thread there. She now allows herself to fall, spinning a thread behind her, to a third attachment point (C). The Y-shaped structure provides the scaffolding for the web. She next builds by turns the radial spokes and frame. She returns to the centre H (which stands for hub, as the centre of the completed web is called), attaches a thread, walks to the outside (B) spinning a thread as she goes, and attaches the other end of it at B; she then walks back down the thread (Figure 2.2c) to a point D where she attaches a new thread. Now, after she has walked on to A, she has constructed a frame from B to A and a radial spoke from D to H (Figure 2.2d,e). The same sequence of movements is reiterated to build the twenty or so radii of the web. In the final web the angles of the radii at the hub are rather constant, at about 15°. The radii are built in a regular order, the next spoke always being added in the largest unfilled sector of the web. In other words, the spider measures the angles between the existing spokes at the hub, and chooses to spin the next spoke between the two spokes subtending the largest angle. Once she has returned to the hub after building a spoke she pulls each

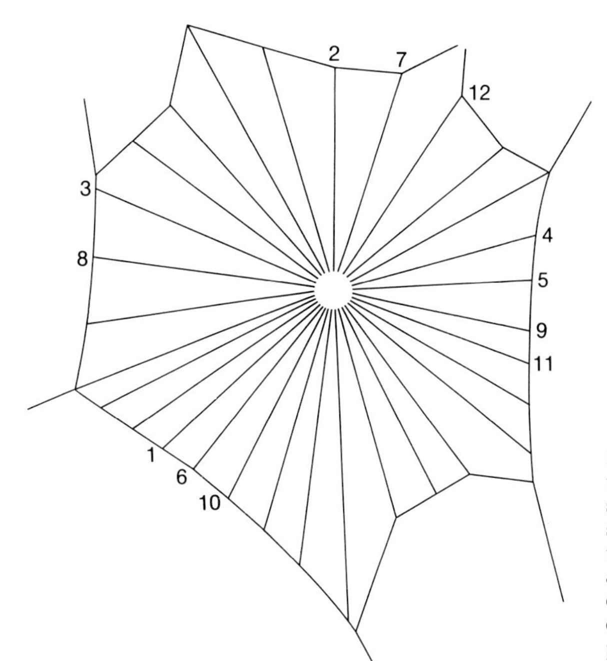

Figure 2.3 An orb web, with most of its radial spokes, but no catching spiral. The numbers indicate the order in which the radii are laid down. Notice that each radius is built in a different sector from the previous one. (After Peters)

spoke with her front legs, which may be how she measures the angles, perhaps by estimating the tension in each spoke. At all events, sight is unnecessary for correct building because spiders can build normal webs in complete darkness, or when sightless. Following the rule for where to spin the next spoke results in spokes being laid down in the kind of order illustrated in Figure 2.3, with each new spoke in a different sector from the previous one.

Once the frame and radii are complete (which takes only about 5 minutes), the spider starts on the spiral. She builds it in two stages. Starting from the hub, she first spirals outwards, to lay down an 'auxiliary spiral'. She does this by walking outwards, in an arithmetic spiral, spinning and attaching a thread behind her, until she reaches the edge. The auxiliary spiral is a temporary structure. For at the outside she turns round and, working inwards now, spins the sticky spiral that will net her prey (Figure 2.4). As she spins the catching spiral she cuts the auxiliary spiral, the function of which was to act as a guideline for laying down the catching spiral. The whole web is finished once the sticky catching thread has spiralled into the hub. Actually, the spiral is not a perfect spiral; it is asymmetrical. Orb webs are not typically circular. They are elongated in the bottom half, with the hub above the centre. The bottom is lengthened by means of a ladder of sticky switch-backs in the bottom sector of the web, which can be seen in Figure 2.1. Only after a few turns of the ladder does the spider begin the true spiral.

The main rules of the orb web spider are first to build a Y-shaped scaffold, then, in a set order, the frame and radial spokes; and finally the auxiliary and sticky spirals. Each main rule contains a set of subrules for measuring angles and walking set distances up certain threads. By following the program of elementary rules the spider can build a complex structure without having a plan of it in her head.

This is not to say that the spider lacks such a plan. Spiders surely are more complex machines than this account has admitted. There is probably more to the control of web building than the series of rules considered here. But this fact does not alter the point of principle which the spider's web illustrates: that apparently complex structures can be built up from a fairly simple set of rules. The principle is of general importance. It is often possible to break down the apparent complexity of animal behaviour by analysing it into a series of 'rules of thumb' that the animal itself could, in some form, be following. Searching for the rules is a powerful method of studying the

Figure 2.4 A partly completed orb web, of *Araneus diadematus*. The auxiliary spiral is complete, but only a few spirals of the sticky catching spiral have been laid down. Notice the catching spiral is much more tightly coiled than the temporary auxiliary spiral. (Photo: Fritz Vollrath)

control of behaviour, because it is easier to pursue research on simple rules than on complex ones. It is easier to follow up the nervous control of a leg movement than of conceptual thought. Of course, complex systems can be studied too; but because it is more difficult it simply takes longer. More sophisticated rules are most easily studied if they are added piecemeal on to a concrete foundation of knowledge. We therefore advance fastest in our understanding of behaviour if we search for the simplest hypothetical mechanisms. Ethologists do not invoke less simple mechanisms — of conceptual thought or conscious calculation, for example — if they are not needed to explain the behaviour that is being observed.

So far we have thought of an animal as executing a series of rules. But this execution must in its turn be controlled by some mechanism. The controlling mechanism ultimately is the nervous system; and to see how that works we must change our study animal, from the araneid spider, to squids and crickets.

2.2 The nervous system

Ramifying throughout the body of an animal is a network of thin white fibres called nerves. A nerve fibre is a bundle of cells called neurons. A muscle contracts when the neuron attached to it, called a motor neuron, becomes electrically active. Neurophysiologists first worked out how neurons work in one particular kind of neuron which is found in the squid. When a squid senses that an enemy is nearby, it suddenly contracts its body, forcing water out of the area called its mantle cavity, and jets away. The squid may also discharge a screen of ink to cover its escape. This 'escape reaction', as it is

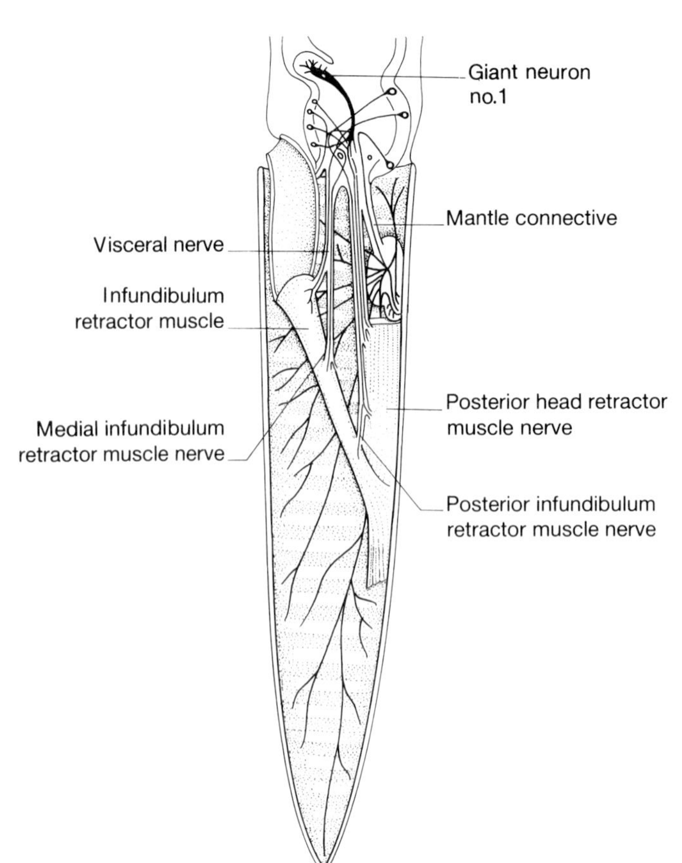

Figure 2.5 The course of the giant neurons of the squid. (After Bullock and Horridge)

called, is controlled by a large (and therefore easily studied) neuron (Figure 2.5), which extends in stages from just outside the squid's brain at one end of its body, to the muscles, the contraction of which at the other end of the body effects the escape reaction.

Microelectrodes stuck inside active neurons revealed that the neuron works electrically. The current is conducted down the neuron by the movement of electrically charged particles (called ions) across the cell's membrane. When the neuron is activated, positively charged sodium ions rush into the neuron. The inside of the neuron in the activated area becomes electrically positive relative to the outside. This pulse of depolarization, as it is called, travels rapidly along the neuron until it reaches the muscle. The electric current then stimulates the muscle to contract. But what stimulated the motor neuron to begin with? We must turn to the other end of the neuron, away from the muscle, where the motor neuron is joined to many other neurons at a junction called a synapse. Here is the origin of motor neuron activity. Chemical neurotransmitters are released by the neuron at the synapse, and if sufficient is released, the motor neuron will fire into action. The motor neuron is thus controlled by the activity of the many neurons preceding it in the nervous system, whose activity is in turn controlled by the central nervous system in the brain.

Sensory input from the external environment is also carried to the brain encoded in electrical depolarizations of neurons, called sensory neurons, which pass from sensory organ to the brain. Speaking very generally, the behaviour of an animal is mainly directed by its brain, which integrates the actions of thousands of neurons, but by processes that are too little understood to be worth dwelling upon.

The neuronal control of some simple behaviour patterns is, however, quite well understood. The singing of crickets is an example, and it will serve to illustrate how the nervous system produces behaviour (Figure 2.6). Singing in crickets is not a response to an external stimulus. It is influenced by external factors such as temperature; but it is not a direct response to them. It is caused by the endogenous action of the nervous system. Male crickets sing, by scraping their wings together, to attract females. The movement of wings is controlled by a set of muscles in the thorax (the thorax is the middle section of the insect, between the head and the abdomen). Motor neurons run from the thoracic ganglion (a ganglion is a mini-brain, an aggregation of many nerves) to the singing muscles. It takes many muscle fibres to move a wing,

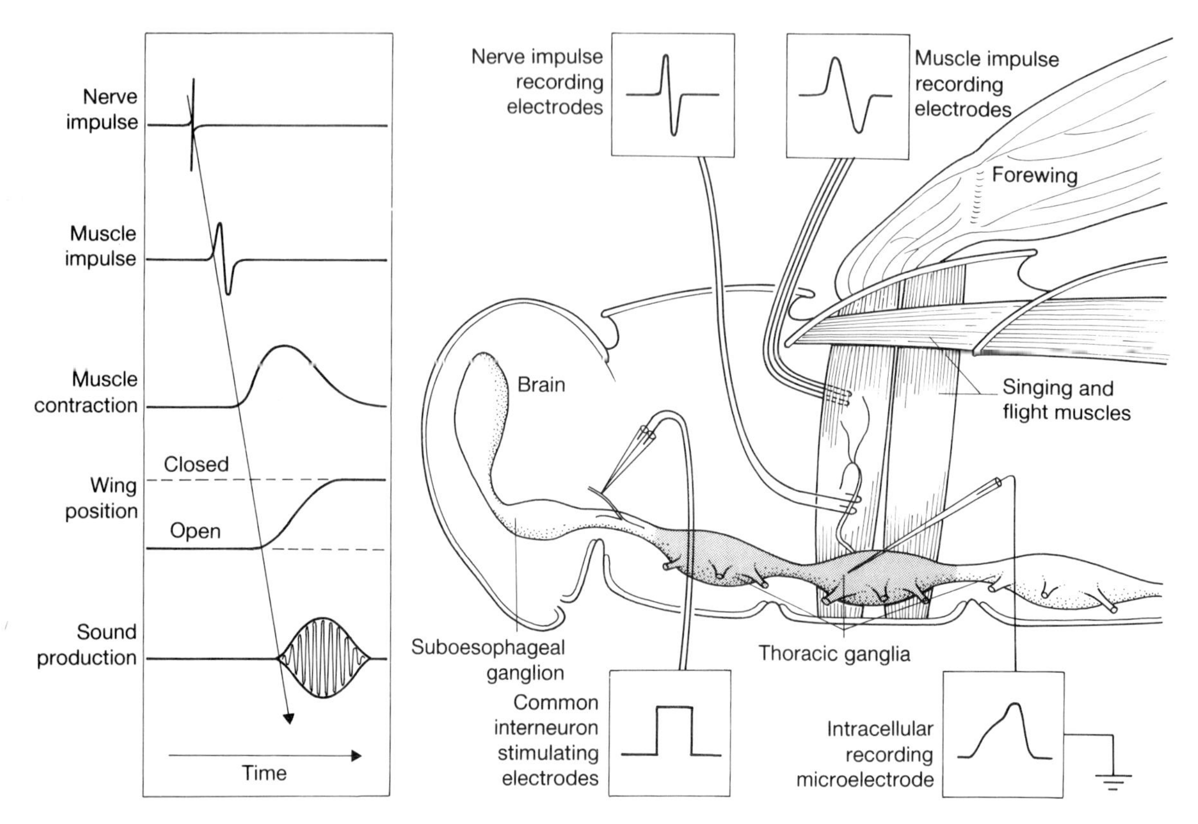

Figure 2.6 The nervous control of the song in male crickets. A normal song can be stimulated electrically in the command interneurons situated between the suboesophageal and thoracic ganglia. The command interneurons control the interneurons of the thoracic ganglia, which in turn control the motor neurons that control the muscles that effect singing, by means of scraping together the wings. (From Bentley and Hoy 'The neurobiology of cricket song'.)

and each muscle fibre has its own motor neuron. All the muscle fibres must contract at the same time, and the co-ordination is produced by a 'command interneuron', onto which all the motor neurons join. Soon after the command interneuron fires, all the motor neurons fire in unison. The co-ordinatory function of the command neuron has been proved by experiment. The command interneuron is situated in the cricket's neck. A neurophysiologist can locate it and join it to an electricity supply. If the electricity is switched on, all the motor neurons fire together, and the cricket produces a perfect song. It will produce its song even if it has had its head cut off. The command neuron, motor neurons, muscles, and wings form a complete singing machine.

2.3 The sensory system

2.3.1 Organs of sense

Animals need to be able to find their way around their environment, find food, recognize the species and sex of other individuals, even (in some cases) whether another individual is a member of the same, or a different, group, and detect the signals sent to them. For all these functions they rely on their sense organs. Different kinds of animals have different sets of sense organs. The set of sense organs possessed by each kind of animal is appropriate to the environment in which it lives. For example, species of fish and shrimp which live in dark, underground caves do not have eyes, or have eyes so reduced that they no longer work; there is no advantage in having light-sensitive organs where there is no light. The human set of eyes, ears, touch, and relatively poor taste and smell is just one particular, not very common, set of sense organs; most species of animals live in a world of senses very different from ours.

We can divide the different sense organs of animals into three groups: exteroceptors, enteroceptors and proprioceptors. This division was suggested by the neurophysiologist Charles Sherrington at the beginning of the century. Exteroceptors sense the state of the environment outside the animal; enteroceptors the state of the animal's body; proprioceptors the animal's movement by sensing the position of its muscles. In mammals, enteroceptors include organs that sense the body temperature and chemoceptors that sense the concentrations of chemicals such as hormones and carbon dioxide. The classic five human senses — sight, hearing, touch, taste, smell — are effected by exteroceptors. In fact there is no perfect classification of these external senses. Taste and smell are both chemical senses (as are many of the enteroceptors). Hearing and touch are both mechanical senses. Other species possess other senses which do not easily fit into the five-way division. But, bearing in mind that the classification is crude, we can consider quickly some examples of each kind. We shall then consider one sense, echolocation in bats, in a bit more detail.

Let us first take a sense lacking in humans, the electrical sense. This has been most studied in fish. Their electrical sense is not the same as the sensitivity to pain by which we become aware of an electric shock, but another sense, comparable to our sense of hearing or smell. Some kinds of fish, such as dogfish (which is a member of the elasmobranch group that includes sharks and rays), use their electric sense to find food buried in the bottom sand by the disturbance to the electric field that the buried living matter causes. The electric sense organs of elasmobranch fish are called the ampullae of Lorenzini and are a set of jelly-filled tubes beneath the skin. Other animals, such as honey-bees, and bacteria, can sense magnetism. Experiments we shall meet later suggest that at least some kinds of birds can also sense the magnetic field.

Chemical sense organs are found in many places in different kinds of animals. The sea hare *Aplysia* (see Figure 2.7), which is a favourite animal for work on the nervous system, can smell seaweed, on which it lives. By recording the activity of the nervous system at different parts of the animal while it is smelling seaweed, it has been found that the chemical sense organs are in its tentacles. Houseflies have chemical receptors in their feet, which enables them to detect food (such as sugary water, in an experiment) by walking into it. Most insects have chemical receptors in their antennae.

The antennae also contain mechanical sense organs; but an insect's whole surface has mechanical sense organs on it. Mechanical sense organs all work by means of tiny hairs. When the hair is bent a sensory neuron attached to it fires into action. Hearing works as a mechanical sense, and is also effected by

Figure 2.7 A compulating pair of seahares *Aplysia*, off the Isle of Mull. They are hermaphrodites and fertilize each other. (Photo: Dick Manuel)

small movement-sensitive hairs connected in some way (depending on the species) to a membrane that is set in oscillation by sound. Fish and some amphibians possess special organs for detecting water pressure called the lateral line organ. The lateral line is a channel under the skin of each side of the animal, with little holes leading to the outside. The flow of water into the lateral line allows the fish to measure the movement of water with respect to itself.

The final class of sense organs is the light receptors. Eyes are found in more or less complex form in many kinds of animals. But eyes are not the only light-sensitive organs known in nature. Insects, for instance, have three light-sensitive ocelli on the top of their heads, behind their compound eyes. The functions of the ocelli are uncertain.

All of the receptors I have just mentioned have been investigated in depth. However, we do not have space for detail about them all. Let us pick on one, the ear, and consider how bats make use of it.

2.3.2 Echolocation in bats

There are over 900 species of bats. The group has perhaps proliferated because its echolocatory sense allows it to live in an environment containing almost no competitors. They fly by night, feeding on night-flying insects, particularly moths. Bats can fly with ease in complete darkness; they do not collide with obstructions, and they catch their prey on the wing. They are capable of flying around a laboratory room which is criss-crossed with a network of wires of a diameter of as little as 3 hundredths of an inch, and catching flying moths from a range of 8 feet. How do bats achieve this? The greatest experimental biologist of the eighteenth century, the Italian Lazzaro Spallanzani, could not solve the problem. He did find that if he stuffed up the bats ears their ability to avoid obstructions declined. But he did not know how to explain this result. Bats remained a puzzle until after advanced equipment for recording and producing sound had been developed along with radar in the Second World War. The equipment was applied to the bat puzzle by Donald Griffin in the 1950s. He proved that the little brown bat finds its way around by listening to the echos of high-pitched sounds that it makes itself.

Humans cannot hear sounds much outside the frequency range 2000–20,000 Hertz (Hz). (The higher the frequency the higher the pitch.) Bats make sounds mainly of 20,000 Hz and higher — some emit sounds of up to

100,000 Hz — and most bat sounds are therefore inaudible to humans. One advantage to the bat of using such high-pitched sounds is that it is undisturbed by background noise (most noises in nature have a frequency lower than 20,000 Hz.) By concentrating on high-frequency sounds, bats live in a world silent except for their own noises. It is essential to the bat that there should be no interfering background noise. In an experiment, Griffin blasted noises of frequencies of more than 20,000 Hz into a room with obstructions. Flying bats now bumped into the obstacles and fell to the floor. As well as showing the importance of a silent background, this experiment also gives part of the evidence that bats use high frequency sounds to find their way around. Echolocation works by measuring the time interval between releasing a short pulse of sound and hearing its echo. The longer it takes the echo to come back, the further away the object must be. The sound pulses have to be very short, to prevent overlap between the emitted sound and its echo: the bat can hear itself as well as the echo. The pulses of sound are indeed very short, the little brown bat releases four or five distinct pulses every second during ordinary flight. It increases the rate of pulses when it detects an object, and as it flies closer to the object its pulse rate rises further. It does so because, as it approaches the object, the delay until the echo comes back becomes shorter and shorter. It therefore makes its pulses shorter, again to prevent overlap between pulse and echo.

Echolocation becomes ineffective beyond distances of about 30–40 metres, because sound is rapidly absorbed in air. But even for these short distances the bat must listen for a very faint echo of its much louder original pulse; the sound pulse emitted by the bat is 2000 times louder than the echo. The bat therefore has the problem that it has to make a loud noise and then hear a soft noise immediately afterwards. When an ear has been blasted with a loud sound it becomes less sensitive for a second or so; human ears, for this reason, take some time to recover after listening to a very loud noise. Bats have a number of methods of solving this problem. One is to make the ear less sensitive when emitting the sound. A nerve going to the muscle of the ear is automatically activated whenever the bat releases its sound pulse. This nerve causes the bat's ear to relax about 5 thousandths of a second before the pulse is emitted, and to recover about 10 thousandths of a second later. The ear will then be at peak sensitivity for picking up the echo.

Bats are not the only kind of animal to use echolocation. They are, however, by far the most extensively studied. Other animals which echolocate are the dolphin and other 'toothed' whales, small mammals called shrews, and a bird which lives in dark caves called the Malayan cave swiflet.

2.4 Putting sensory information to use

Granted that an animal possesses various kinds of sensory information, how can it use them to control its behaviour? This is part of a much wider question, because animal behaviour is not only a set of simple responses to various environmental situations; but we can make some points about how an animal uses its sensory information without implying that it is all there is to behaviour. All sense organs send their knowledge in coded form to the central nervous system by means of a sensory neuron. Different sensory neurons are stimulated by different properties of the environment. Light sensitive neurons, for instance, contain a light sensitive pigment, which, when illuminated, changes its chemical form and causes the neuron to burst into action. A special neuron is required to sense light: if you shone light on any other kind of neuron it would have no effect on it; only a neuron containing a light sensitive pigment is set in action by light.

Most behaviour patterns are controlled in the central nervous system which integrates the information from the sense with other neuron systems, to control the animal's behavioural output. Some behaviour patterns are not centrally controlled. They are called peripheral reflexes (of which the human knee jerk is an example), and are controlled by a simple system of one sensory neuron and one motor neuron. The sensory neuron connects directly on to the motor neuron, which in turn controls the muscles that effect the behaviour pattern. When the sensory neuron fires (after mechanical stimulation in the case of the knee jerk reflex), it stimulates the motor neuron, which causes certain muscles to contract. We shall meet some other examples of reflexes; but in most behaviour patterns the sensory input and behavioural output of the animal are less directly connected, and the senses exert their influence on behaviour through the central nervous system.

There is potentially an enormous amount of information about the environment, far more than an animal could make use of in deciding on a course of action. The environment, for instance, is brim-full of electromagnetic rays: it has a constantly changing pattern of light and dark, different colours, and a continual hum of X-rays and radio waves, together with high and low frequency radiation of which we are not normally aware.

Most of the information is, for any given behavioural decision of the animal, completely irrelevant. To avoid walking into a tree you only need to know where its edges are; the fine detail of colour patterns on the trunk and leaves do not matter. To behave appropriately, therefore, an animal has to select its information, and that is exactly what it does. Some of the selection is done in the sense organ itself, which will only sense certain patterns; and the rest of the selection is carried out as the information is centrally integrated into the nervous system.

Given that animals are selecting from the available information in their environment, how can we find out what they are actually responding to? One method is neurophysiology. We can record the activity of sensory neurons when an animal is successively presented with a variety of objects. We shall meet one example of this kind of study in the prey-catching behaviour of the toad (p. 106). The same kind of question can also be tackled by a grosser method. We can ignore the physiological intermediary details, and find out what kind of environmental stimuli the animal responds to at the behavioural level.

The simplest kind of sensory information is used in the responses called 'kinesis' and 'taxis'. In a kinetic response, the animal alters its rate of movement, in a random direction, according to the intensity of the stimulus. When the stimulus, which might be light, or moisture, is of the right intensity it slows down and thus spends more of its time under those conditions. Woodlice show a kinetic response to moisture. They move faster where it is

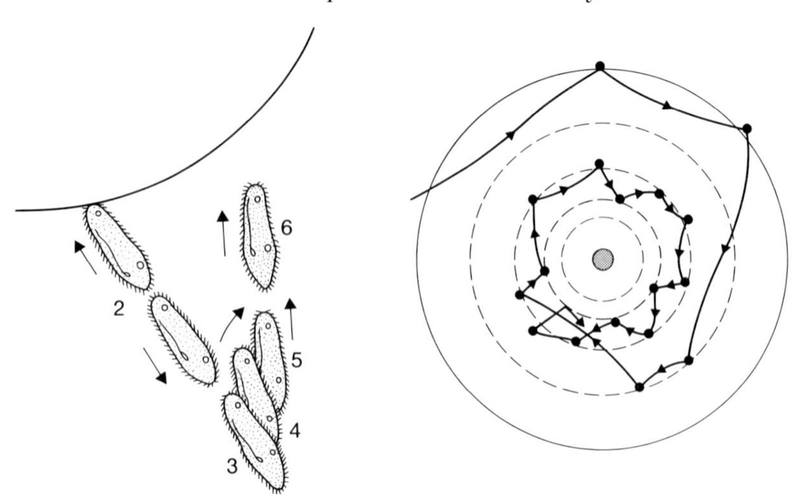

Figure 2.8 Response of the unicellular organism *Paramecium* in the vicinity of a CO_2 bubble. When it senses a high CO_2 concentration it withdraws, turns through a certain angle, and advances again; but its new direction is unrelated to the direction of the stimulus. (After Kuhn)

drier, and therefore spend more time where it is moist. The unicellular organism called *Paramecium* shows a kinetic response with respect to the local concentration of carbon dioxide (Figure 2.8). A taxic response, however, is directional. Negative phototaxis, for example, means that the animal moves away from light, as in fact does the maggot of the bluebottle fly. The taxic response is made by sensing the direction of the light, which is achieved by different techniques in different species. It requires only the most elementary kind of sensory information. The animal does not need to know anything about the light source, only that it is light and where it is coming from.

Few behaviour patterns, except perhaps chemical responses to food (or pheromones, see p. 127), are guided simply by the intensity of one stimulus. More of the behaviour patterns that we can see animals performing are controlled also by the pattern of the stimulus in the environment. For example, in the case of light, the exact distribution of light and shade, which defines the shape of the image, would be important, rather than only the presence or absence of light. Animals must be able to recognize patterns in the environment. We saw, for instance, in the last chapter, that a goose will respond to an egg outside its nest by rolling it in. Gulls do likewise. But what exactly is it that stimulates the behaviour? How does a gull recognize an egg? Gerard Baerends and his colleagues sought to answer this question by experiments with model eggs of various appearance. They made model eggs, which varied in their size, stippling, shape and colour. They presented herring gulls, on their nests, with choices of different model eggs, and recorded which ones were preferentially retrieved (Figure 2.9). The gulls

Figure 2.9 A black-headed gull with a choice of an artificially large egg (of natural colour and stippling) and a natural egg. The gull is retrieving the large egg. (Photo: Niko Tinbergen)

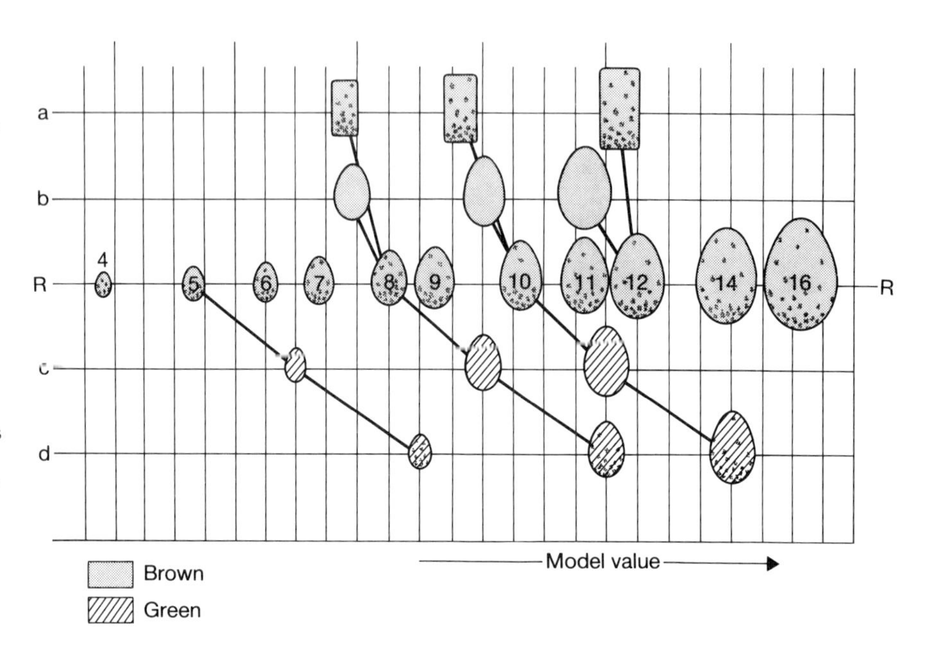

Figure 2.10 Preferences of herring gulls for different model eggs. Herring gulls were offered, for retrieval, a choice between a model egg of a certain size, shape, colour and stippling, and another egg of natural appearance but not necessarily of natural size. Gulls prefer to retrieve larger eggs, and a scale of increasing egg size can be used as a scale to measure how the gulls assess other properties of the egg. The scale of increasing sized eggs is the R-series in the middle of the Figure; the number in the egg indicates its size relative to natural size in units of ⅛ (8 is natural size, 16 twice natural). There are four experimental series: (a) brown, stippled, block-shaped; (b) brown, unstippled, egg-shaped; (c) green, unstippled, egg-shaped; (d) green, stippled, egg-shaped. The preference for a model egg can be read off the graph by comparing its position on the horizontal axis with an egg of equal size in the R-series. For instance, a-series eggs (block-shaped) are displaced a little to the left of the R-series, which suggests that herring gulls take little notice of the shape of the eggs. Eggs of equal size are connected by thick lines for ease of comparison. (After Baerends and Kruijt)

preferentially retrieve larger models, even if their size exceeds the natural size of an egg, and the size of an egg can therefore be used as a scale of comparison for the other variables. Baerends tested each model egg by giving gulls a choice between the model egg (which might be varied in its colour, shape or stippling) and a range of sizes of control models (of normal shape, colour and stippling). Figure 2.10 summarizes the results. The gulls take more notice of some variables than of others. They take little notice of shape: an oblong model is as likely to be retrieved as an egg-shaped model of the same size (compare *a* with *R* in Figure 2.10). Nor do the gulls prefer eggs of natural coloration: they preferentially retrieve a green egg rather than a naturally coloured egg of the same size (compare *c* with *R*); but they are sensitive to colour, as they prefer green to brown eggs (compare *b* with *c*). The other variable, besides size, which strongly influences the gulls' preference is stippling; they prefer stippled to unstippled eggs (compare *d* with *c*). It is as if natural colour and shape are not part of a gull's idea of an egg, but size and stippling are. They do not simply prefer to retrieve eggs of natural appearance, and indeed prefer larger than normal eggs, in which case the models take on the condition of a 'supernormal stimulus': the model is a more attractive stimulus as the stimulatory property is exaggerated. The supernormal stimulus in a sense deceives the egg recognition mechanism of the gull. In nature, the mechanism works to distinguish eggs from other objects; but it can be tricked by experiment. However, the main conclusion from Baerends' experiment is that herring gulls recognize eggs mainly by the criteria of size and stippling.

'Behavioural assays', such as the egg retrieval response of birds, not only reveal what stimulus pattern is recognized by the animal; they are also a revelatory method of studying the sensory powers of animals. If an animal can be shown, by appropriately controlled experiments, to behave in response to some property of the environment, it must be able to sense it. Karl von Frisch applied the method to demonstrate the hearing ability and colour sensitivity of fish. In that case, the physiologist von Hess had asserted that fish are colour blind and deaf; von Frisch doubted the assertion, and he successfully trained minnows to distinguish colours by rewarding them with food, and catfish to come out of a tube when he blew a whistle. In both experiments he used a behavioural response to discover a sensory ability.

2.5 The control of behaviour

2.5.1 Choices among behaviour patterns

The behavioural output of an animal emerges as a sequence of many different behaviour patterns, and each change in the sequence can be thought of as a behavioural 'choice' made by the animal. The question is how animals make those choices. Some will be responses to changed environmental stimuli. New sensory information is one factor causing an animal to choose one behaviour pattern rather than another, but it is not the only one. The same animal may not respond to the same stimulus in the same way on different occasions, and it may change what it is doing even while the environment appears to be constant. There are two, related reasons why an animal may behave differently when under similar environmental conditions. One is that its internal tendency to behave in a certain way may change. For example, when food is presented, it will become less likely to feed as it grows less hungry. The other reason is the interaction of behavioural preferences; an animal may stop feeding in order to avoid a predator. The two reasons are related because the internal tendency of an animal to behave in a certain way is presumably determined by a balancing act among the consequences of all its possible responses to a given environment. The balancing act, and the resultant behavioural preferences or tendencies, are called the 'motivation' of the animal. Let us consider how the motivation of an animal influences its behavioural choices.

A study by Baerends, on the freshwater fish called guppies (*Poecilia reticulata*), provides a clear example of the interaction of motivation and external stimulus. The behaviour in question is the courtship of females by males. Baerends recognized three different behaviour patterns that a male may perform when courting a female: 'posturing' in front of a female, a limited sigmoid movement, and a full sigmoid display. How does a male decide which to perform? The answer seems to depend on the size of the female and the male's own motivation to court, which can be independently measured by his coloration. Figure 2.11 depicts the combinations of these two factors necessary for a male to court a female in each of the three ways. The particular shape of the graphs is not important here; they are only to illustrate a general point, which is that, for any behaviour pattern in any species there

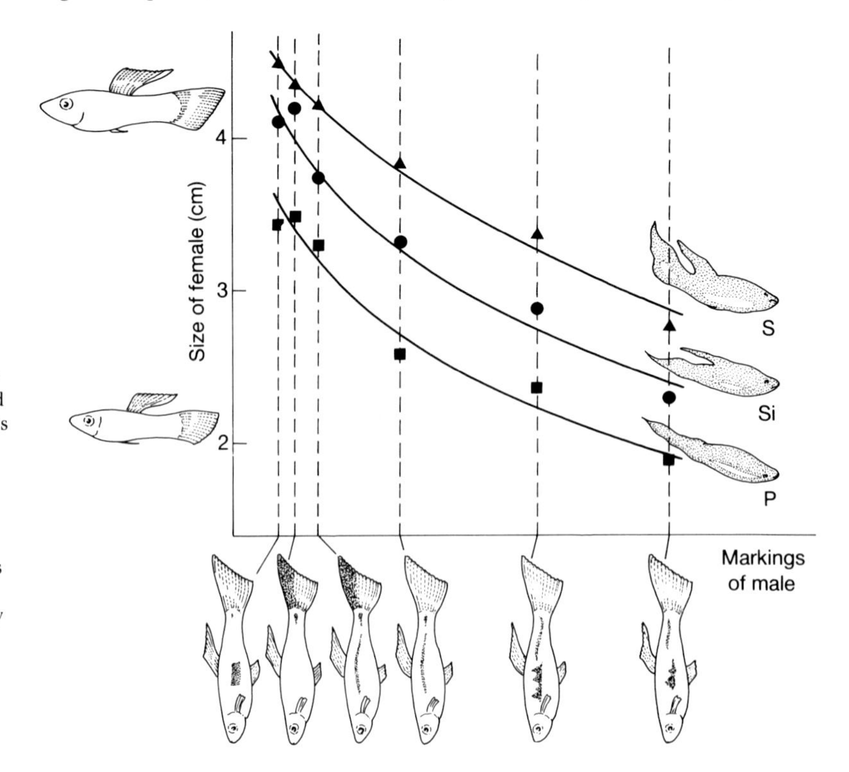

Figure 2.11 The tendency of male guppies to court females is determined by an internal factor, how inclined the male is to court (which is indicated by his colour markings), and an external factor, the size of the female. The curves connect points where males will perform each of three kinds of courtship display, in order of increasing intensity of courtship: posturing (P), sigmoid intention movements (Si), and full sigmoid displays (S). (After Baerends *et al.*)

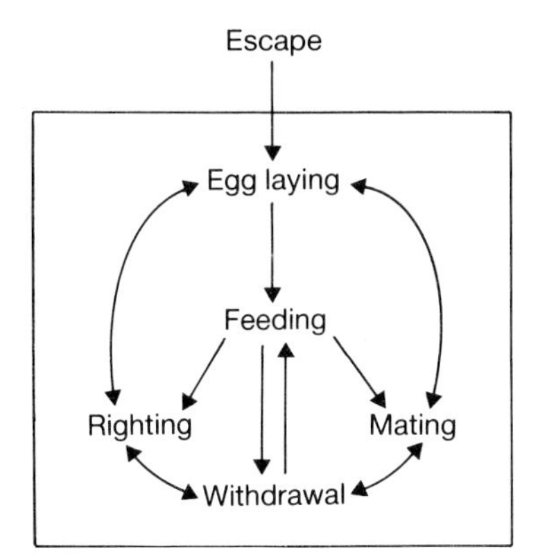

A → B A dominates B

A ↔ B A and B can be simultaneous

A ⇄ B Reciprocal inhibition between A and B

Figure 2.12 The behavioural priorities of the intertidal snail *Pleurobranchia*. 'Escape', for instance, is performed in preference to 'egg laying' when the circumstances appropriate to both behaviour patterns are fulfilled. (After Kovac and Davis)

will be some such graph of motivational tendency and external stimulus which describes the conditions under which it is performed.

The guppy illustrates choice among different behaviour patterns of one class, courtship. What of interactions among different kinds of behavioural goal? Here we need a new example, which (unlike the courtship of guppies) is understood neurophysiologically. Actually, little progress has been made in the neurophysiological study of behavioural choices. Nervous analysis is difficult enough for single behaviour units, let alone interactions among many activities. The study of the gastropod *Pleurobranchia* by J. W. Davis and his colleagues has, however, partly uncovered the neurophysiological control of six behaviour patterns. The six are: feeding, egg laying, escape, withdrawal of the oral veil, righting, and mating (Figure 2.12). Take first the interaction of feeding and egg laying. *Pleurobranchia* is a carnivorous snail which includes eggs in its diet. When a *Pleurobranchia* lays its own eggs, it switches off its feeding habit. Another behaviour pattern, escaping from predators, is performed in preference to all other activities. A fourth behaviour pattern is to withdraw its oral veil on being touched. A snail with its oral veil withdrawn cannot feed, and its tendency to withdraw its veil interacts with its tendency to feed. If food is abundant, or the snail is not hungry, withdrawal has priority over feeding, and *vice versa* when food is scarce and the snail is hungry. The other two activities studied by Davis are 'righting' (turning the right way up) and mating. Having established the behavioural priorities by observation, he proceeded to their neurophysiology. The system has not been completely elucidated, but it is now known for instance that two neurons are responsible for inhibiting the 'withdrawal' response when a *Pleurobranchia* is feeding, and that the inhibition of feeding during egg laying is effected hormonally. The priorities of *Pleurobranchia*, by the way, do make sense, for without them it would eat its own eggs after laying them; and if escaping from danger did not have absolute priority, it would not survive to exercise its other behavioural preferences.

2.5.2 Hormones

The nervous system controls the behaviour of animals over a short time scale, of matters of seconds or micro-seconds. Other factors control it over longer terms of days, months, or even years; the most important of these are hormones. Some hormones act quickly; but we shall concentrate here on those with relatively slow effect.

Hormones are chemicals that circulate in the bloodstream of animals, regulating the animal's metabolism and behaviour. They are released by special glands, such as the pituitary gland at the base of the brain, and the gonads (the ovaries in the female, the testes in the male). The glands release their hormones into the blood; some other organ will then respond to the increased level of hormone circulating in the blood. The responsive (or target) organ is often, but not always, the nervous system. An example of a target organ other than the nervous system is the effect of the hormone testosterone in the African clawed toad. Testosterone is released by the testes. An increase in the amount of testosterone in the clawed toad's blood stimulates the growth of special 'nuptial pads' on the male's front legs, which the male uses to embrace the female while mating.

The most comprehensive study of the hormonal control of behaviour concerns the reproductive cycle of the Barbary dove (Figure 2.13). It was worked out by Daniel Lehrman and his colleagues. The reproductive cycle lasts about six to seven weeks. Before the beginning of the reproductive season, the level of testosterone in the male's blood is low. Male Barbary doves with low levels of testosterone are aggressive to females. The aggression of the male in turn suppresses the release of reproductive hormones in the female, which ensures that she does not become ready to reproduce before the male. The reproductive season comes on as daylength increases; its

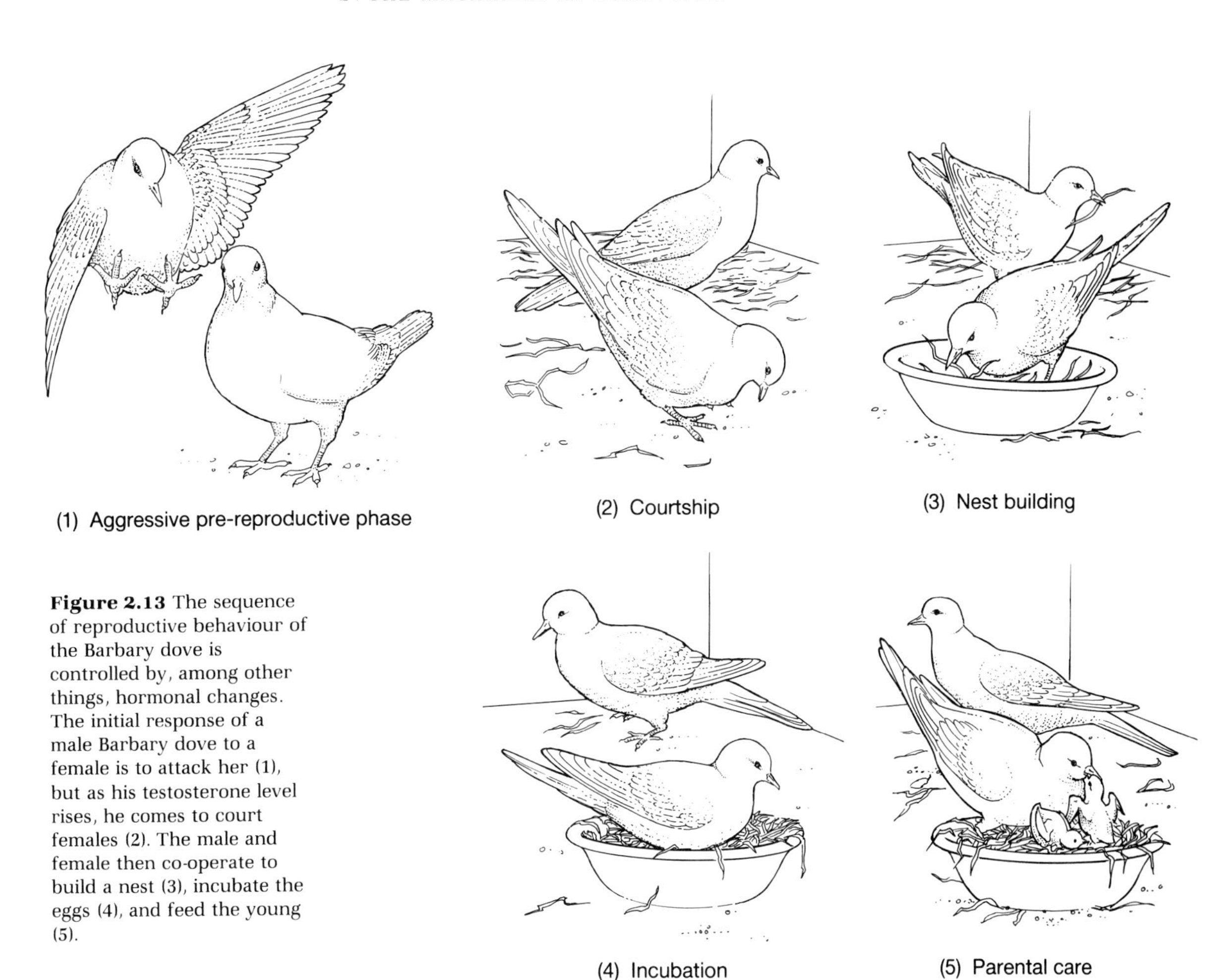

Figure 2.13 The sequence of reproductive behaviour of the Barbary dove is controlled by, among other things, hormonal changes. The initial response of a male Barbary dove to a female is to attack her (1), but as his testosterone level rises, he comes to court females (2). The male and female then co-operate to build a nest (3), incubate the eggs (4), and feed the young (5).

beginning is determined hormonally, as the increase in the number of hours of daylight stimulates the male's testes to release testosterone. The testosterone acts in the Barbary dove's brain. In the male, it causes him to cease being aggressive to the female, and to start courting, which consists of a ceremony of bows and coos. That testosterone is responsible for the change can be show experimentally by injecting it into a male who will soon start courting, if given a female. Males injected with another hormone, oestrogen, show the same response, because testosterone is converted into oestrogen in the brain before it exerts its effect of inducing courtship. Courtship is therefore stimulated by testosterone released from the male's testes, but after it has been converted into oestrogen on the way.

The bowing and cooing of the male Barbary dove influences the hormonal system of the female. It stimulates her pituitary gland to release a hormone called gonadotropin, which stimulates her ovaries to grow, in preparation for egg production. The stimulated ovaries in turn release oestrogen, which stimulates the female to start building a nest. After a few days the nest is built and the female then lays two eggs. Meanwhile, the nest building will have stimulated, in both male and female, the release of the hormone progesterone, which brings on the next stage of the reproductive cycle — sitting on the eggs. This stimulates the release of yet more progesterone, and of another hormone called prolactin. Prolactin causes the changes in the male and female needed to make them able to feed their young. For Barbary doves feed their young on a special kind of 'milk'. Dove 'milk' is made in the crop (a region of the gut between the mouth and the stomach); bits of the lining of the crop break off, and are later regurgitated to the young as 'milk'.

The behavioural changes through the reproductive cycle of the Barbary dove are thus controlled by a series of five hormones, testosterone, oestrogen, gonadotropin, progesterone and prolactin. Each hormone stimulates the next stage of behaviour, which in turn stimulates the release of the next hormone. But the release of hormones is not only controlled by the behaviour of the animal releasing them; the hormonal system is also influenced by the external environment (as in the effect of daylight on the male) and the behaviour of other animals (as in the effect of male aggression on the female). Many other examples of the hormonal control of behaviour are known; but we can allow this particularly comprehensive study to stand for them, for it illustrates the main principle of the mechanism.

2.6 Summary

Behaviour is controlled by the nervous system, which operates electrically and chemically. Motor neurons attach to muscles, and stimulate their contraction. In the case of the muscles that effect singing in the male cricket, different muscle fibres are coordinated by a 'command interneuron', which ensures that all the motor neurons leading into a muscle fire synchronously. Animals inform themselves about their external (and internal) environments through their sense organs, the kinds of which vary greatly among different animals, in ways appropriate to their different environments. The question of what aspects of their environment animals respond to has been studied as profitably through their behaviour as their neurophysiology. Senses which were doubted, such as hearing in fish, and unsuspected ones, such as echolocation in bats, have been discovered by behavioural techniques. But the sequences of behaviour produced by animals are not controlled only by external stimuli; they also depend on internally-determined preferences and the animal's motivation. It is easy to say that an animal's choice of what to do is made according to both internal motives and external stimuli, and the point can be illustrated by the courtship of male guppies, but there is no general, powerful theory of the control of sequences of behaviour. Likewise, little is known about the neurophysiological control, although we do possess one good analysis in the snail *Pleurobranchia*, whose choices among escaping, egg laying, and feeding are effected by a combination of neuronal and hormonal mechanisms. Hormonal influences on behaviour are normally longer-term than, and often mediated by, neuronal mechanisms. Much is known about hormonal control of behaviour; the principle is simply that hormones are stimulated to be released by one organ, they cause a change at another, to result in a behavioural change. The well-understood hormonal control of the reproductive cycle of the Barbary dove, from courtship, to nest building, to egg laying, to incubation, to lactation (of a sort), illustrates the principle.

2.7 Further reading

Camhi (1984) introduces and thoroughly reviews 'neuroethology', and his Chapter 3 is a clear introduction to the nervous system in general. Bentley and Hoy (1974) explain the mechanism of cricket song; Bentley and Konishi (1978) is a more advanced review. See Schmidt (1978) for sensory physiology; and Sales and Pye (1974) for echolocation. The early work on the control of behaviour, and on the use of models to study pattern recognition is still as easily read about in Tinbergen's classic *The Study of Instinct* (1951) as anywhere. Halliday (1983) introduces the science of motivation, and McCleery (1983) discusses Davis's work on *Pleurobranchia* in the context of an introduction to behavioural decision making. Lehrman (1964) explains his work on the hormonal control of reproductive behaviour in Barbary doves, and Slater (1978) reviews the subject more generally.

3 / The genetics and development of behaviour

3.1 The principles of genetics

Inheritance has been the subject of some of the most important scientific discoveries of the past century. The mechanism of inheritance was cracked using non-behavioural traits, but we can reasonably infer that behaviour is inherited in much the same way as are other, more thoroughly studied properties of organisms. It will be easiest to explain the principles of heredity with the traits in which they were originally discovered, and only then to turn to behaviour. For behaviour has both been less studied than some parts of morphology, and its inheritance is generally too complex to lay bare the elementary principles.

Genetics is now an advanced science, but we only need consider its elements. The first important discovery was made by Mendel, in the mid-nineteenth century. He experimented on pea plants. He first isolated pure lines of eight paired traits; of which we shall confine ourselves to the lines of 'tall' and 'short' peas. (A pure line is one that breeds true for that character; when two peas from a pure 'tall' line are crossed, they always produce a tall pea.) Mendel crossed tall peas with short peas and found that the first generation were always tall. He then crossed together members of the first generation, and found that in the second generation tall and short peas occurred in the ratio 3 tall: 1 short. The explanation of Mendel's result is as follows (Figure 3.1). There are two factors, which we call genes. One causes tallness, and may be called *A*; the other causes shortness, and may be called *a*. Each pea has two such genes. Peas with *AA* are pure tall and peas with *aa* are short. (*AA* and *aa* were the two starting pure lines.) When peas are crossed each pea puts only one of its two genes into its offspring. Because there are two parents, the offspring receives two genes controlling size (one from each parent). All the first generation peas in Mendel's experiment received one *A* from their tall parent and one *a* from their short parent. They were, therefore, *Aa*. All these peas were tall. This is explained by *A* being 'dominant' to *a*: if an organism contains two genes, one dominant to the other, then it develops the character controlled by the dominant gene. (The other, unexpressed gene is called recessive.) Now consider what happens when two

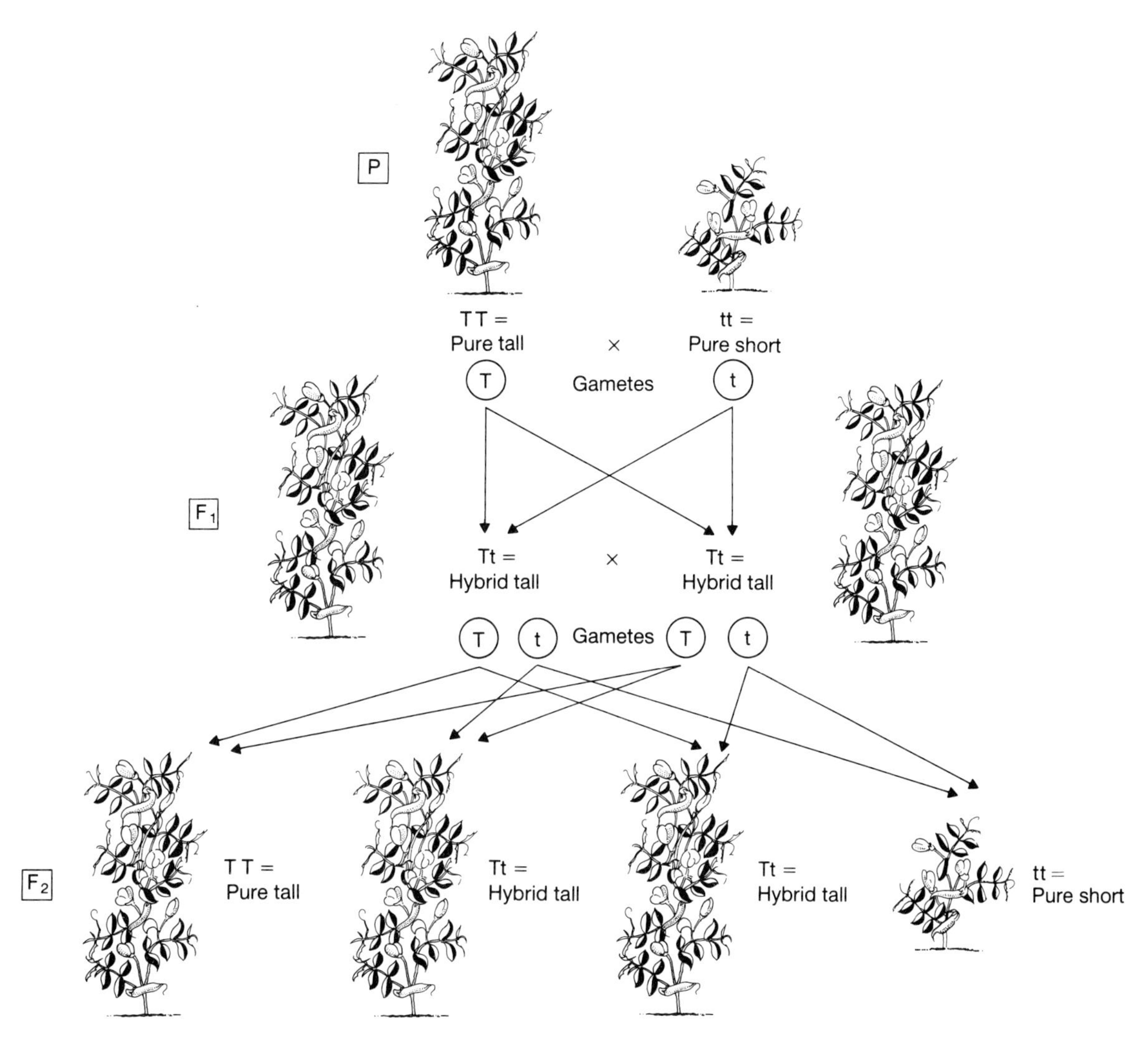

Figure 3.1 Mendel, in one of his original experiments on heredity, crossed a pure line of *tall* peas with a pure line of *short* peas. The first generation of peas were all tall. He then crossed members of the first generation; and in the second generation the peas were tall and short in the ratio 3 to 1. He explained the result as follows. Each pea has two factors controlling height, but only one is passed on to its offspring. Each factor may be of one of two kinds, causing either tallness or shortness. The *tall* pure parental line had two tallness factors, the *short* line two shortness factors. Their offspring had one of each but were tall rather than intermediate in height because the tallness factor is 'dominant' to the shortness factor. When the hybrid first generation were crossed, the tallness and shortness factors, in all possible combinations, produce ratios of three tall plants to each short one. The Mendelian factors are now called genes.

of the first generation tall peas are crossed. They are both *Aa*, and their genes can combine in four different ways. An *A* from one parent can combine with an *A* (giving *AA*) or an *a* (giving *Aa*) from the second parent, and the *a* from the first can combine with an *A* (giving *Aa*) or an *a* (giving *aa*) from the second. *AA*, *Aa*, and *aa* are therefore produced in the ratios 1:2:1. However, *Aa* peas look like *AA* peas — they are both tall — so the ratio of tall: short is 3:1.

Inheritance, then, is effected by paired units, called genes, which determine the characters of the organism. The genes *A* and *a* that control size in peas are just one of many genetic 'loci' in peas. (A locus is simply the location of a gene on a chromosome.) Each organism contains thousands of genes; a human for instance, possesses perhaps between 10,000 and 50,000 genes, each with its individual effect on the development of the individual. In technical language, the difference between the genes that an organism inherits and the actual characters of the organism is the difference between its genotype and its phenotype. The genotype is the set of genes it inherited, the phenotype is what it looks like. The means by which genes produce their effects are known in some detail. Genes themselves are inherited in long chains called chromosomes, made of the chemical DNA. (Chromosomes can be seen with the aid of a microscope.) Genes directly dictate the manufacture of molecules called proteins, and it is through their proteins that genes exert their effects on the phenotype of the body. Although we know much about how genes work at the microscopic level, less is known about all the intermediate processes through which genes find expression in phenotypic characters at the macroscopic level of the whole animal. We possess no worked example to illustrate how genes control behaviour; all we can say is that, in abstract terms, proteins influence the development of neurons, and that genes must by this means influence the development of the nervous system. That, however, is a vague remark; and must remain as such, because the mechanism by which genes build bodies is one of the great unsolved problems of biology.

Size in peas, in the form Mendel experimented on it, was an exceptional trait; most traits, particularly behavioural ones, do not come in simple discrete pairs. Many traits, such as size in humans, vary continuously, or at least come in multiple forms (i.e. humans are not either short or tall, but a wide range of heights). Mendel in fact selected the paired categories of size because he reasoned, correctly, that the mechanisms of inheritance would be most easily revealed in such a trait. Continuously varying traits differ from

Mendel's peas in that their variation is probably controlled by a large number of genes, and by environmental differences too.

Inheritance can be studied in continuously varying traits, but slightly different techniques are needed — the techniques of quantitative genetics. For a continuously varying trait (let us use human size as an example), the value of the trait in an individual is probably determined by what genes it has at a large number of genetic loci, together with the effect of the environment. In the case of Mendel's peas size was controlled by two forms of one gene (each such pair of genes at one locus, is termed an allele, various possible alleles can occur at one locus but there will only be one pair per individual). In humans size may be controlled by multiple alleles at many dozens, or hundreds, of loci. (I should stress that I am using human size as an imaginary example: I do not know how many genetic loci and alleles are really at work in this case.) The whole genotype controlling size in an individual would then be impossibly difficult to work out by Mendelian breeding experiments. Imagine for instance a case where size was influenced by 22 loci, each with only two alleles (*A*/*a*; *B*/*b*; *C*/*c*. . .*T*/*t*; *U*/*u*); and that each capital letter allele added 3 inches to the individual's height, each lower case allele added nothing. An individual with two capital letter alleles at all 22 loci would then be $2 \times 3 \times 22$ inches = 11 foot tall. An individual (in this imaginary example) with lower case genes at all 22 loci would have zero height. The genetic difficulty concerns all the intermediate heights. Take, for example, an individual who is 5½ feet tall. His genotype could be *AaBbCcDd*. . .*Uu*; but it could also be *AABBCC*. . .*JJKKLLmm*. . .*ttuu*; or any other combination of genes such that half were capital forms and half lower case forms. These are a large number of possibilities and they would be practically impossible to distinguish by breeding experiments. In a Mendelian experiment on pea size we can find out what genotype an individual has, whether it is *AA*, *Aa*, or *aa*. For a trait influenced by more than a small number of genes we cannot work out the genotype by a Mendelian experiment. If we are to study the genetics of quantitative characters we should aim to know less than the exact genotype responsible for each phenotype.

In practice, although the exact genotype of an individual cannot be found out, we can find out something more abstract if less informative. We can find out the 'heritability' of the trait. Heritability is a number between 0 and 1, which expresses the extent to which the variation of the trait is due to variation in genotypes and in environments. In terms of size in humans, some differences between individuals will not be caused by differences in their genes at all, but by differences in how they grew up, how much they ate and so on; other differences will be genetic. Thus, if we take the population as a whole we can ask what proportion of the variation in size is due to genetic variation and what proportion to environmental variation. Heritability is defined as the proportion due to genes; it is the ratio of genetic variation to the total variation. Examples of the two extremes may clarify the meaning. Heritability is zero if all the variation is non-genetic. For example differences in the languages different humans speak, are probably entirely caused by a non-genetic factor; in this case the language spoken by those around us when we are young. The heritability of human language is probably zero. At the other end of the scale, heritability is one if all the differences among individuals are due to their genes. Few traits reach the extreme of a heritability of one, although blood groups in humans are an example.

The important point about heritability is that we do not need to know anything about the actual genotypes in order to say what it is. The heritability of a character is independent of the genetic details. It is a more abstract, cruder kind of knowledge. We could calculate the heritability for a character of which we did know the exact genetics (such as Mendel's peas), but there would be little point in doing so, because the actual genotypes controlling the trait are more informative than the heritability. Measuring heritability comes into its own for those characters for which we have no knowledge of the genetics.

So far I have only said what heritability is. How can it be measured? The fundamental technique is still the controlled breeding experiment. There are two important forms of breeding experiment. One is to breed together different parental lines that differ for the trait of interest. If we wish to know the heritability of size we breed together sets of parents of different sizes, and measure the size of the offspring. If there is a genetic influence, the offspring will, on average, resemble their parents. In practice we should draw a graph of the parental values against their offspring values, and if the graph shows any relationship the character is heritable. The strength of the relationship indicates the degree of heritability; if there is no relationship at all, heritability = 0 (figure 3.2). (The method works with other classes of relatives as well as parents and offspring. If a trait is heritable, all classes of relatives will be correlated to some extent.)

The second technique is the artificial breeding experiment. Artificial

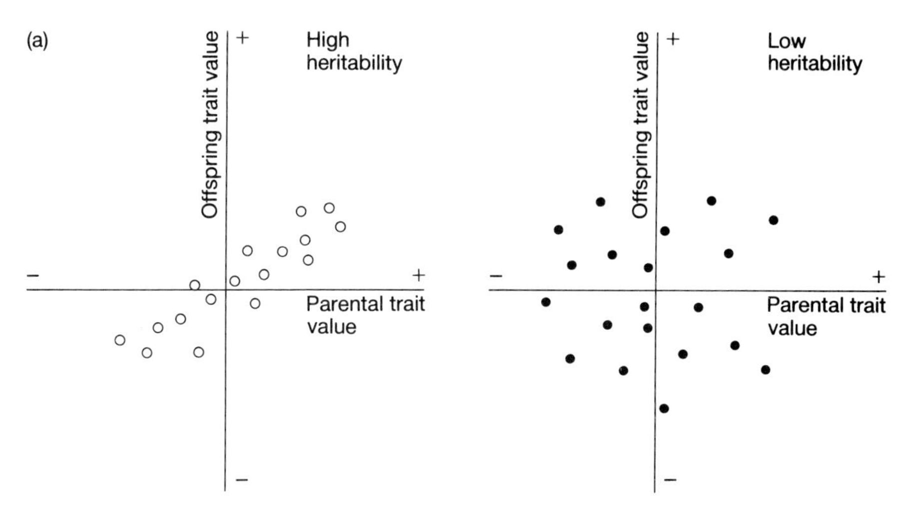

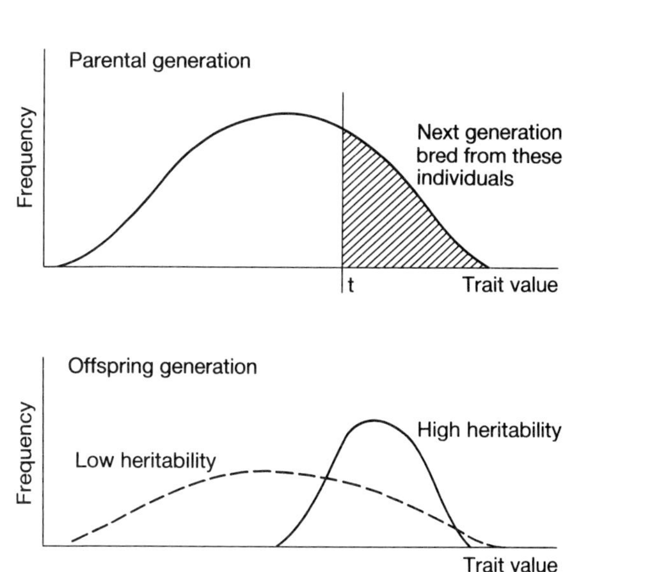

Figure 3.2 Measurement of heritability. (a) by breeding experiment. The value of the trait (which might be some such quantitative variable as size) is measured in the offspring and their parents: if the values are correlated then (provided that the offspring have not been reared in a similar environment to their parents) the trait is heritable. (b) by artificial selection. Only those organisms of more than some value (t) of the trait are allowed to breed. If the average value of the trait in the offspring is higher than the parental generation average, the trait is heritable.

selection means breeding only from a selected minority of the members of a population. It is frequently used in agriculture. If we wish to increase growth rate in pigs, for example, we can try breeding only from the fastest growing pigs, and a whole population of fast growing pigs may be obtained in a few generations, if growth rate is heritable. If artificial selection does change the average condition of the population, the trait under selection must have been heritable; and the degree of response to artifical selection indicates the degree of heritability. If fast growth is not heritable, it would be impossible to produce a line of fast growing pigs by selective breeding.

The heritability of a continuously varying trait is a useful statistic in the absence of an exact knowledge of its genetics. It tells how much of the variation among individuals in a population is due to variation in their genes; which in turn tells us whether the population would respond to natural selection on the trait. (For natural selection only works on heritable traits, as we have seen.) If we are trying to understand why different animals behave differently, the value of the heritability of the behaviour in question will tell us whether one possible factor — genetic variation — is at work.

Now that we have established the principles of inheritance, and the methods by which it can be studied in complex cases, let us see how well they apply to behaviour. We shall consider two examples; one in which the Mendelian principle appears to apply in simple form and another in which a genetic influence on behaviour has been demonstrated by artificial selection. The genetic control of behaviour patterns in different species is, for all we know, so diverse that neither experiment should be thought of as particularly representative of behaviour as a whole. They only show how it can be studied in some cases.

3.2 Behavioural genetics

The first example concerns the honey-bee, and we need to know a little about its habits. Honey-bees lives in hives in large societies of about 15,000 bees. In nature, bee hives may be built in any convenient hollow, such as the inside of an old tree. The hive consists of a series of vertical, wax honeycombs in which the bees store their food and rear their offspring. The offspring live in the combs while they are eggs, larvae, and pupae. They then come out as adults, ready to contribute to the work of the hive. Most bee hives nowadays are made by humans, and the vertical shelves on which the bees build their combs are built so that they can be slid in and out of the hive.

Honey-bee larvae are susceptible to disease. One disease in particular, called American foul brood, is caused by the bacterium *Bacillus larvae*. A larva which catches foul brood dies. The dead, rotting larvae then becomes a source of infection to any other larvae in nearby combs. Some strains of bees prevent the spread of the disease by removing the rotting larvae. Such strains are called 'hygienic'. The hygienic bee performs two acts. It removes the cap of the cell containing a rotting larva, and it throws the larva away. Not all bee strains are hygienic; W. C. Rothenbuhler, therefore, could experiment on the inheritance of the habit and showed that the difference between hygienic and non-hygienic strains is due to their possession of different genes.

First Rothenbuhler experimentally infected combs of larvae with foul brood. He then slid these combs into different hives, and simple observation revealed which hives contained hygienic bees. He then crossed a hygienic strain with a non-hygienic strain, and tested their hybrid progeny with combs of infected larvae. All the hybrids were non-hygienic (Figure 3.3). The genes for hygienic behaviour therefore must be recessive. In the next stage, Rothenbuhler did a 'backcross' of the hybrids with the hygienic parental strain. He reared 20 colonies from such backcrosses in all. Six of the 29 were

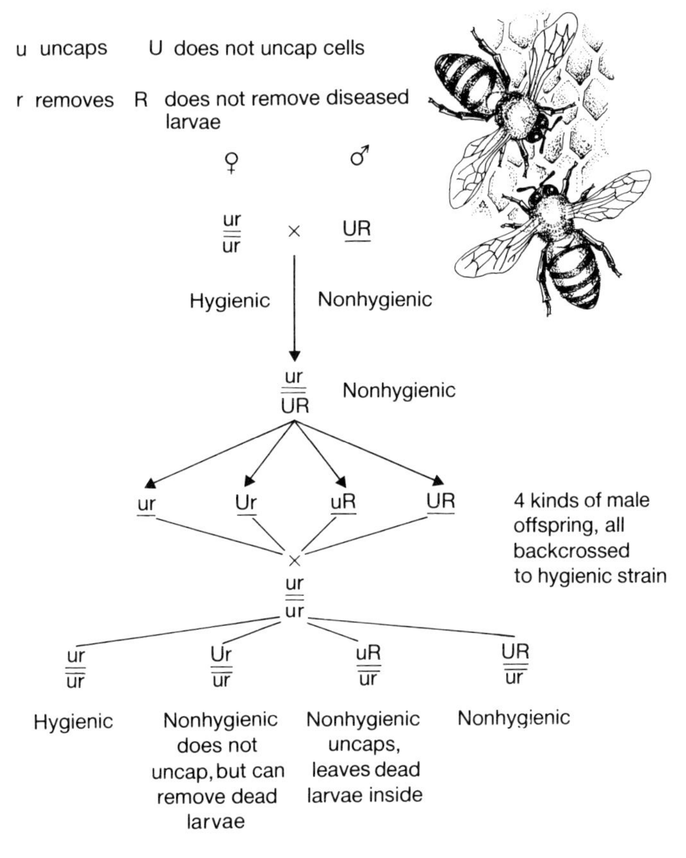

Figure 3.3 Rothenbuhler studied the genetics of hygienic behaviour in honey-bees by crossing a pure unhygienic strain with a pure hygienic strain. His results are explained by supposing that there are a pair of genes, each of two kinds. One gene controls the uncapping of the brood chamber; the other the removal of the dead developing bee larva. If this interpretation is correct it should be possible to breed four different kinds of honey-bee: fully hygienic (remove chamber caps and diseased larva); remove chamber cap but not diseased larva inside; remove larva but not cap; fully unhygienic (remove neither). From the cross of the two pure strains, and experimental removal of caps in unhygienic hives, Rothenbuhler found all four kinds.

hygienic (6/29 is approximately one quarter). About another quarter (9/29) behaved very strangely. Bees of these nine colonies took the caps off the larval cells, but then left the dead larvae inside. The remaining half (14/29) of the colonies were non-hygienic — they left cap and larva alone. Rothenbuhler thought that hygienic behaviour might be controlled by two recessive genes. One gene would control uncapping; the other the throwing out of the larvae. The nine colonies which uncapped but left the dead larvae inside had one of the genes, that for uncapping, but not the other for removing. Rothenbuhler therefore reasoned that, of the 14 colonies which were apparently nonhygienic, about seven might have the gene for removing the dead larvae. Because these seven lacked the gene for uncapping, the gene for removing would not be expressed. Accordingly, he removed the caps from the combs of foul brood-ridden larvae. He put these combs into his 14 hives. As he had predicted, six of the colonies promptly removed the dead larvae.

Rothenbuhler's experiments on the hygienic behaviour of bees provides a very clear example of the genetic control of behaviour. Each of the two behavioural stages of hygiene is controlled by one gene. It is likely that most behaviour patterns are controlled by many more than two genes. Hundreds of genes probably control most behaviour patterns. A behaviour pattern controlled by hundreds of genes does not give such clear categories as those found by Rothenbuhler. In Rothenbuhler's experiment there were two genes, with two alleles each, and four (2^2) categories of bee. If there were one hundred genes, each with two alleles, there would be 2^{100} categories; the experiment would therefore be much more complicated.

When animals do not fall into a few simple categories, quantitative genetic techniques become more practical. Majerus, O'Donald, and Weir, for instance, were interested by the genetical control of mate selection in ladybirds. Most ladybird populations contain only one type, the familiar beetle with small black spots on red background, this type is called *typica*. In some populations in Britain, however, there are also ladybirds with more black on their wing covers (the black spots are enlarged and more numerous) this type is called *quadrimaculata*. They had noticed that females of both types generally prefer to mate with a *quadrimaculata* male, if given a choice. However, not all females show equally strong preferences, and they decided to breed from the females that did mate with a *quadrimaculata* male, to see whether their offspring would also show the preference. They continued the experiment for four generations, through which the strength of the

Table 3.1. Artificial selection of mating preference in the ladybird *Adalia bipunctata*. The preference measures the degree to which females mate with a *quadrimaculata* male when given a choice between *quadrimaculata* and *typica*; if the preference were zero, mating would be random, if one, all matings would be with *quadrimaculata*. (Simplified from Majerus *et al* 1982)

Experimental line		Control line	
Generation	Preference	Generation	Preference
1	0.18	1	0.27
2	0.39	2	0.27
3	0.46	3	0.29
4	0.57	4	0.21

preference tripled (Table 3.1). Evidently the variation in the mating preference is due in part to genetics.

3.3 Development

When discussing the genetics of a trait, it is easiest to think of the trait as constant within an individual. The principles apply equally to inconstant traits, but are more clumsy to express. In fact, of course, many behavioural traits are not constant throughout the life of an individual; they are subject to developmental change. How does behaviour develop? We cannot make generalizations here like those for inheritance, for development is less well understood. Individual questions about behavioural development have however been answered, and we can look for patterns in them.

Many studies of development consider what factors an individual must experience in order to acquire a particular behaviour pattern. Consider, for instance, the song of the male cricket: do the males learn the song by listening to other males? The crucial experiment is to rear male crickets without allowing them to hear the song of other crickets. If crickets learn their song, the experimentally isolated males should not be able to sing a cricket song. In fact they can; learning is unnecessary. Notice that the experiment only rules out one (or a few) experimental factors. It shows that the sound of other males' singing is not necessary for the development. It does not show that *no* experience of any sort is necessary, as indeed it could not, for it is logically impossible to substantiate universally negative statements. That, however, does not prevent us from drawing particular, limited conclusions from such 'isolation' experiments.

In crickets, the ability to sing can develop without the experience of song, but in many birds the story is not the same. In most species of birds, a male reared apart from other singing males will not develop a proper song at all; this is not true of simple bird sounds, such as the cock crow, but no 'song birds' are known to be able to develop their elaborate songs in isolation. If, however, the isolated bird is played a tape recorded song of its own species, it will later be able to sing it normally. We shall return to this subject. Another case in which experience has been shown to influence development is the pecking of chicks. Gull chicks, for instance, peck at their parents' bills, which causes them to regurgitate food. Newborn chicks will peck at an adult gull's bill — or even a model of one — without any prior experience. Jack Hailman found that a herring gull chick will initially peck equally at a model of a herring gull adult, or of a laughing gull (which is quite different, its head being mainly black, unlike the white herring gull). After a few days in the nest, however, the chicks will peck more at the herring gull model (Figure 3.4). They have experienced being fed by herring gulls, and have modified their preference accordingly, in a way that Hailman calls 'perceptual sharpening'. Experience here influences development.

Actually, this experiment alone does not demonstrate that experience is necessary: the change could take place automatically with age. We should distinguish learning from 'maturation'. Maturation means that the change in the behaviour is due to other changes in the animal, not to practice of the behaviour itself. Chick pecking again provides an example, but this time from pecking food on the ground, not at a parent's bill. As soon as a domestic hen chick hatches it starts pecking at grains that look like food, and as it grows

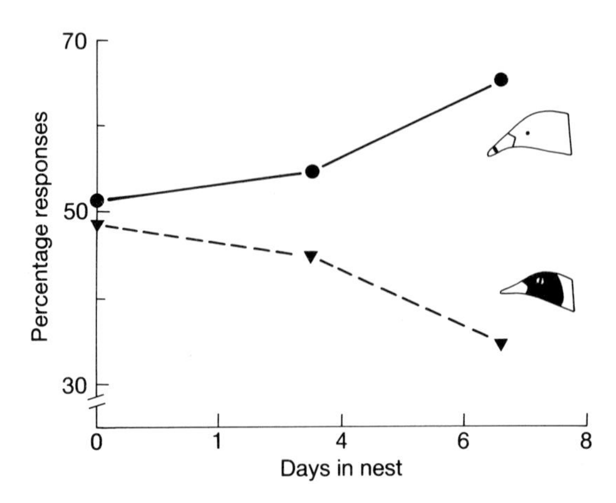

Figure 3.4 At birth, herring gull chicks will peck equally at models of adult bills of either herring or laughing gulls; but after seven days they have developed a preference for the model of the adult of their own species. (After Hailman)

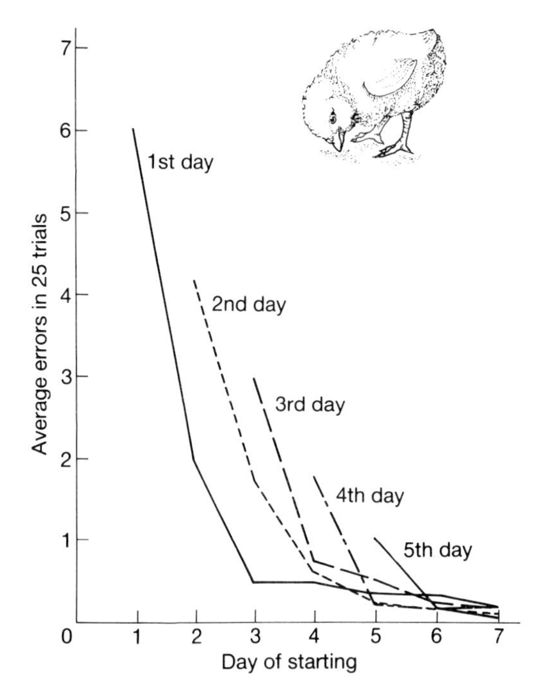

Figure 3.5 Chicks feed by pecking at small grains of food. Their pecking is not perfectly accurate to begin with, but improves with time. These graphs show that there are two components in the improvement. One is learning: the error rate decreases with practice: each line illustrates a decrease in the error rate since starting. But there are five lines, for different classes of chicks allowed to start to peck from one to five days after hatching. The initial error rate decreases from one to five days, which indicates that the improvement within a line is due to maturation as well as learning. Chicks started on their third day, for instance, made fewer errors to begin with than chicks started on their first day. Even without practice, the accuracy of pecking improves. (After Cruze)

older its aim at food grains improves. This is in part due to practice; but there is also an influence of maturation, for if a chick is prevented from pecking at food during its second day, it will still be better at pecking on its third than on its first day (Figure 3.5) but it will not be as accurate as it would have been if it had been allowed to practise. Accurate pecking in chicks therefore develops, by processes both of maturation and learning, from an initially inaccurate condition to one of greater accuracy.

We have considered whether an experience of particular factors might be necessary for the development of given behaviour patterns. Let us now make the question more precise, and ask whether the experience has the same effect at any time, or whether there are particular 'sensitive periods' at which it will be more influential than at others. Most research on sensitive periods has been conducted in relation to the phenomenon called imprinting; it will therefore be convenient to take the two together.

3.4 Imprinting

If you remove the eggs of a goose from their mother, you can hatch them in an incubator. When the goslings hatch, they will recognize the first moving object they see as the creature to follow around for the next few weeks. If it is a human that they see first, they will follow him or her (Figure 3.6). It helps if the human makes some distinctive noise as well, which they can then recognize by sound as well as sight. The goslings will have become 'imprinted' on the human. In nature, of course, the first moving, honking thing that a gosling sees is its parent, and it will normally become imprinted on, and follow, its mother. It is only when intrusive ethologists steal and hatch eggs that the wide tolerance of the goslings is revealed. Goslings will not only imprint themselves on human beings or geese, but also on inanimate objects such as a cardboard box, a rubber ball, or even a flashing light. Young birds, however, are not completely undiscriminating; they do have preferences. Pat Bateson has found, for instance, that inexperienced one-day-old domestic chicks prefer a red flashing light to a yellow one, if given the choice. Why they do is not clear, but the preference exists. However if a chick is first imprinted on a yellow flashing light, it will then prefer it to a red one.

After the young animal has imprinted itself on a particular individual, its attachments are fairly irreversible. It will continue to be attached to that

Figure 3.6 Goslings will 'imprint' on a human foster-parent if that is the individual they see after hatching. These goslings have imprinted upon Konrad Lorenz. (Photo: Niko Tinbergen)

individual even after the period of parental care is over, not changing its attachments to some other animal. A lamb will become imprinted on whoever feeds it. If that individual is not its mother but a human foster parent feeding it from a bottle, the lamb imprints on whoever held the bottle. Even after the lamb has been weaned and has joined a flock it remembers the hand that fed it. If its human foster parent comes nearby, the lamb will leave its flock and stand near its foster parent.

Filial imprinting takes place in many species of birds and mammals. These are the kinds of animals with most extensive parental care. Imprinting is adaptive because it enables the young to recognize and follow their parents. They will grow up in a world of many hostile enemies and one or two protective parents. If they are to survive, it is important that the young should choose the right animals to follow. The kind of imprinting that we have been considering so far is called 'filial' imprinting. It is the imprinting of the following response which young animals make to their parents. Other kinds of behavioural response can also become imprinted. Sexual imprinting, for example, concerns the species to which the animal will direct its sexual behaviour. In geese, sexual imprinting and filial imprinting are two different processes. A goose which follows a human being round as if it were its parent does not have its sexual behaviour anything like so disturbed; when it grows up it will court other geese. Another kind of imprinting may take place in young salmon smolt. According to one hypothesis (p. 90–91), the smell of its home stream is memorized by the young salmon, and when it grows up it will migrate back to the river that smells like its home stream. If this is so, the migratory response is imprinted in salmon.

Imprinting, as I remarked earlier, illustrates the principle of 'sensitive periods'; it is usually established during a specific period early in the animal's life. The exact timing of the sensitive period differs between species; domestic chicks, for example, only follow objects they have seen during the first three days after hatching, whereas for mallard ducklings, the phase lasts for 10–15 days after hatching. The chick will not imprint on objects seen after that time.

Sexual imprinting also occurs in early life. Most experiments on sexual imprinting have been done on birds. It has been found that birds are most easily sexually imprinted on their own species, fairly easily on closely related species, and only with difficulty on very different species. Herring gulls and lesser black-backed gulls are similar species, and black-backed gulls reared by herring gull parents, because some ethologist moves eggs between nests, become sexually imprinted on the herring gull: the adult lesser black-backed gulls so produced will try to mate with herring gulls. They are sometimes successful. Most of the gulls which are hybrids between the two species around the British Isles are the result of these experiments. Sexual imprinting normally functions in the wild to ensure that the animal will, when it grows up, choose a mate of the correct species. In nature, to look at your parents is a good method of learning the characteristics of your own species.

3.5 Learning and memory

Learning can, in a serviceable but imperfect definition, be said to include any change in an individual's behaviour that is due to its experience. A convenient distinction, which helps to organize the research on learning, is that between non-associative and associative learning. In associative learning, the animal learns that different properties of the environment — different stimuli, as the properties are called — are associated, and modifies its behavioural responses to one of them accordingly. For example, a chimp may learn an association between poking a stick into termite mounds and extracting a stick covered with edible termites. It will then learn to poke sticks in termite mounds when it is hungry. In non-associative learning the animal also learns to modify its behaviour but not because of any association of stimuli. Concrete discussion should make the distinction clear.

3.5.1 Non-associative learning: habituation and sensitization

Two main kinds of non-associative learning are recognized, called habituation and sensitization. We can take our examples from the sea hare *Aplysia* (see p. 36). *Aplysia* breathes through its gills, which are situated in a region called the mantle cavity; the gill's enclosure opens to the outside through an opening called the siphon. If an experimenter prods the siphon, the *Aplysia* withdraws siphon and gills, and folds them up within the mantle cavity. This is called the gill withdrawal reflex, and is simply a protective reaction. After a while, if undisturbed, the *Aplysia* puts its siphon out again, and if it is then prodded a second time, it will show the same withdrawal reflex. However, it will not do so an indefinite number of times. If it is repeatedly prodded, it comes to ignore the stimulus, and leave its siphon and gills out. This is the kind of behavioural change called habituation — the *Aplysia* has learned not to respond to an apparently harmless stimulus. Sensitization is the opposite kind of change. Habituation means to become less sensitive to a stimulus, sensitization more so. If an *Aplysia* receives an alarming stimulus such as an electric shock on the tail, it then responds more readily to other stimuli (such as prods to the siphon) that it would otherwise have been less responsive to. It has become more sensitive, as indeed makes sense, for if an *Aplysia* receives a dangerous stimulus naturally, it probably means some hazardous entity is nearby, and it will pay to be careful.

I picked on the *Aplysia* examples, because they lead to one of those rare cases in which we have a neurophysiological understanding of behaviour (Fig. 3.7). The nervous control of the gill withdrawal reflex is a simple unit of one sensory neuron and one motor neuron. The siphon contains the sensitive end of the sensory neuron, which, at its other end is directly connected, at a synapse, with a motor neuron controlling the muscles of the mantle cavity. When the sensory neuron is stimulated, it fires the motor neuron, and the siphon and gills are withdrawn. How does the system habituate? There are two possible mechanisms for regulation in so simple a system: either a change in the amount of neurotransmitter released by the sensory neuron, or a change in the sensitivity of the motor neuron to constant doses of neurotransmitter. In the case of *Aplysia* the former possibility is the real one. It is as if repeated activity in the sensory neuron exhausts its supply of neurotransmitter, making the system as a whole less responsive.

Two more kinds of neurons are needed for sensitization (Figure 3.7). The dangerous stimulus is sensed by another sensory neuron, whose synapse connects with one or more interneurons that eventually run into the same synapse connecting with the motor neuron that controls the mantle muscles. When the sensory neuron, for instance in the tail, becomes active, it fires the interneuron, which in turn causes a series of chemical changes in the motor neuron. The effect of those chemical changes, the details of which are known, is to make the motor neuron fire more readily. It requires less of a stimulus to become depolarized. Hence the system is sensitized.

Habituation and sensitization are not the only kinds of non-associative learning: imprinting, which we discussed in an earlier section, is another kind; and the development of bird song, which we shall discuss later, yet another. It has been a matter of controversy whether the non-associative kinds of learning are really learning at all, and if so, whether they do not really take place by the same general mechanisms as associative learning. It is

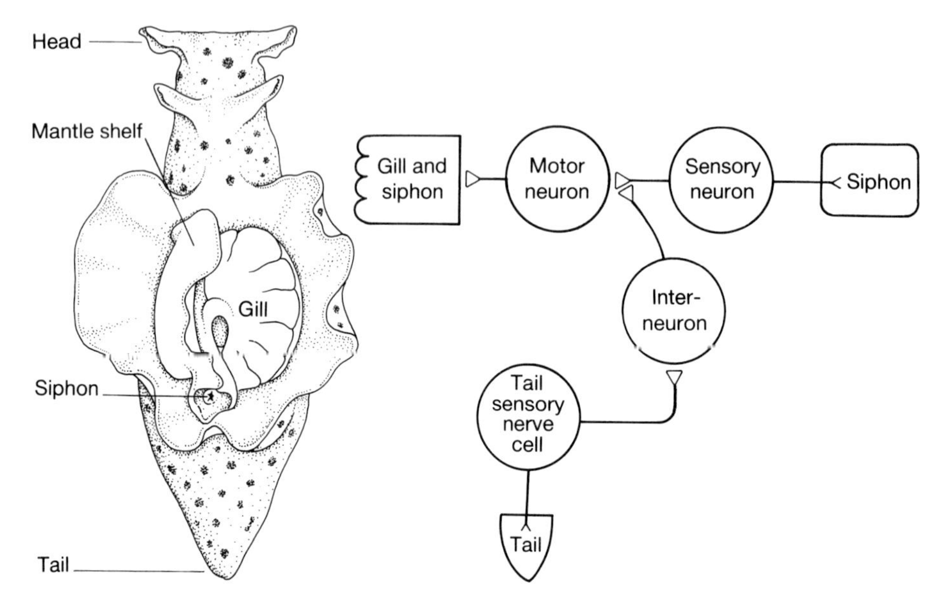

Figure 3.7 The nervous control of siphon withdrawal in the sea hare *Aplysia*. When an *Aplysia* is tapped on the siphon, it withdraws its siphon and gills into its mantle chamber; but if it is repeatedly so tapped, it ceases to respond. This 'habituation' is controlled in the synapse of the siphon's sensory neuron and motor neuron.

likewise a matter of unsettled controversy how far the kinds of neuronal mechanisms identified in *Aplysia*'s non-associative learning could apply to learning in general. Comparably clear facts are not available for other systems, and as we move on to associative learning we must return from the neuronal to the behavioural level. Most of the research on learning has been carried out on unnatural behaviour patterns, in the laboratory, particularly in two species: rats and pigeons. The kind of associative learning shown by rats and pigeons in these experiments is often called conditioning. Let us see what this means.

3.5.2 Associative learning: classical and operant conditioning

There are two main kinds of conditioning, called classical conditioning and operant conditioning. We shall take them in turn. Classical conditioning was first studied in dogs by the Russian physiologist Ivan Pavlov (1849–1936). Pavlov's interest was in digestion. In 1904 he won the Nobel Prize for his work on the physiology of digestion, after which he turned to study the conditioning of digestion. In a typical experiment Pavlov would sound a bell when bringing a dog its food. As the dog learned the association between the sound of the bell and being fed, it salivated on hearing the bell in expectation of its meal. Soon Pavlov could make his dog salivate just by sounding the bell,

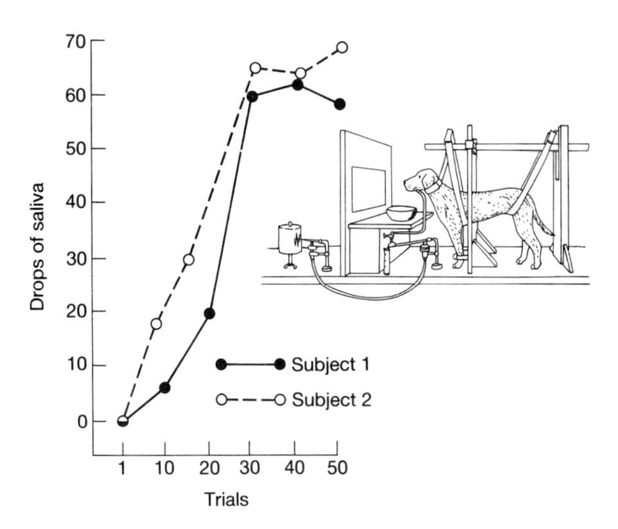

Figure 3.8 Classical conditioning of salivation in dogs. The dogs come to salivate on hearing the sound of bell, which they associate with being fed. The data were obtained as follows. The trial number of the horizontal axis indicates the number of times the dog had been fed and heard an associated sound. After various numbers of trials the dog had its salivation measured after it had heard the sound, even though it was not fed on that occasion. (After Anrep)

even without bringing its food (Figure 3.8). He claimed the salivation of the dog had been conditioned. Classical, Pavlovian conditioning is useful to the animal when it needs to be able to anticipate some repeatedly occurring change in its environment. An animal might for instance prepare itself for a fight by learning a particular association between some noise and the appearance of its enemy in the near future.

The other main kind of conditioning is called 'operant' or 'instrumental' conditioning. It was first worked on at the turn of the century by the American psychologist, E. L. Thorndike, and more recently by another American psychologist, B. F. Skinner, and many others. The experimental apparatus they used has come to be called a Skinner box. The pigeon in the box has a choice of two coloured disks to peck at. If it pecks at one it receives a grain of food; if it pecks at the other it does not. There are endless possibilities for further experiments like this. One can make one disk give food at one rate, and the other at a slower rate. One can see what happens if the animal sometimes receives food on pecking a disk, and sometimes does not. One can vary the time between the pecking and the food being presented. In any case, the bird learns the association between doing something and being fed, and accordingly it more or less accurately pecks the correct disc when it is hungry. The important point is that the animal learns an association between its behaviour (pecking) and a consequence of the behaviour (being fed), and modifies its behaviour appropriately.

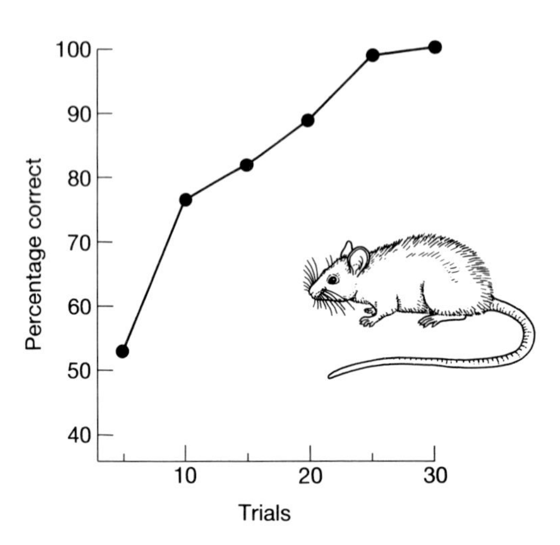

Figure 3.9 Operant conditioning of maze running in rats. The rats have to turn in a certain direction in a T-maze in order not to be electrically shocked or in order to be fed. They choose the correct direction with increasing accuracy as the number of trials (the number of times they have been allowed to run through the maze) increases. (After Clayton)

Analogous experiments have demonstrated associative learning in rats. Instead of pecking at discs, rats can be taught to press levers in order to obtain food and water. They can also be taught to run through mazes. The simplest maze is a T-maze in which the individual has to make one directional choice; by rewarding rats that turn one way rather than the other (or by giving them an electric shock if they turn in the opposite direction), the experimenter can teach them to make a consistent and predictable choice. The results (e.g. Figure 3.9) can be plotted in a graph of the rate of correct choices against the number of times the rat has been made to run through the maze (where a 'correct' choice is one in the direction in which it will be rewarded, or in the opposite direction from which it will be shocked). The rate of correct choices increases with the number of trials, as the rat learns to modify its behaviour (its direction of turning) as a consequence of the behaviour itself (the consequence being food or an electric shock). Again, it has learned an association.

3.5.3 How chaffinches learn chaffinch songs

Associations with bells or illuminated disks are rather artificial, but they are intended to be experimental versions of things that would be important in nature. Ethologists have studied other more natural kinds of learning, such as the development of bird song. We have already seen that male song birds, isolated from the sounds of their own species, do not develop a normal song;

but considerably more is known concerning what auditory experiences are needed, at what time, in order for the bird to learn to sing properly.

The chaffinch (*Fringilla coelobs*) has been particularly well studied. There seem to be two important stages, one of memorizing, the other of practising, its song. The memory phase lasts from hatching to about ninety days later. Chaffinches do not sing at this time, but memorize the sound of adult

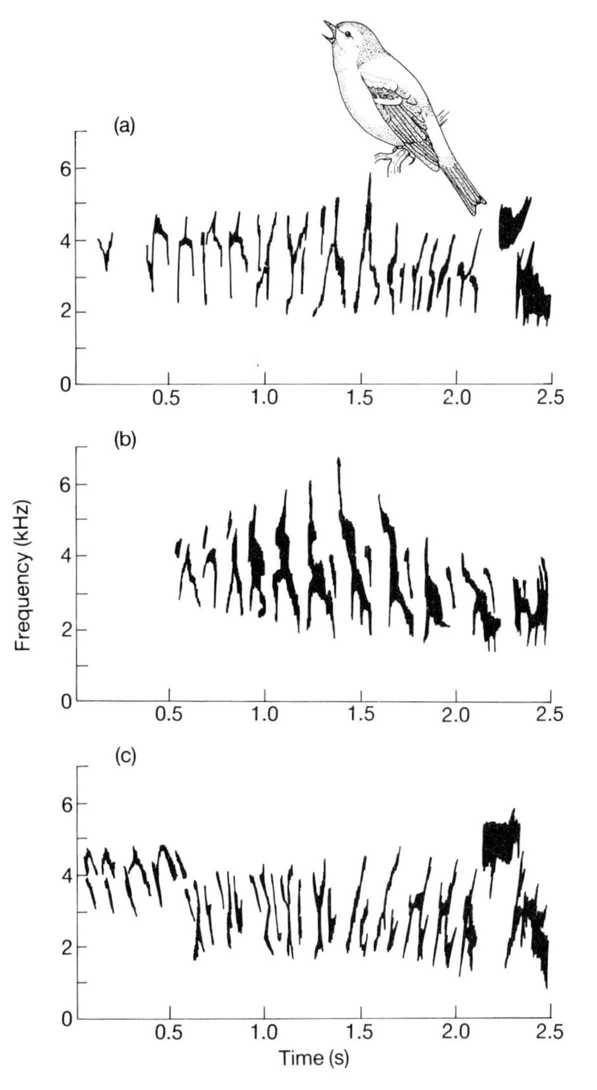

Figure 3.10 Sounds are shown visually by sonograms, which illustrate the frequencies of sounds being made through time. The sonogram (a) at the top is a normal chaffinch song. (b) In the middle is the song of a chaffinch that had been reared in complete isolation, unable to hear any sounds at all; it resembles the normal song, but is a crude version of it. The chaffinch must have opportunity to learn if it is to sing the normal song rather than the crude version. (c) At the bottom is the song of a young male, sung at the beginning of his first spring. He gradually improves it with practice into the normal chaffinch song shown at the top. (After Thorpe)

chaffinches. A chaffinch that is isolated from birth will only sing a very crude chaffinch song when it comes to sing in its first year (Figure 3.10b — compare the normal chaffinch song in Figure 3.10a). However, young chaffinches that have heard chaffinch song during their first 90 days also sing a poor version of it in their first spring (Figure 3.10c); but they improve with practice until they sing a normal song. Chaffinches reared normally for their first 90 days but then deafened sing the customary crude song (Figure 3.10c) at the beginning of their first spring, but they do not improve. It has therefore been suggested that, during their first 90 days, the chaffinches memorize normal chaffinch song, and form a 'template' inside their heads with which (unless deafened) they compare their own attempts. They then, with practice, bring their own song into line with the memorized template. The development of bird song therefore has an abstract similarity with the development of chick pecking which we discussed earlier. The behaviour initially appears in a crude form, and then, with practice and in some cases maturation, it improves into the normal form of adults of its species. Sensitivity is confined to a special phase.

The development of chaffinch song is most obviously classified as non-associative learning, because the bird learns its song by comparing it with a template rather than because of any consequence of singing a better chaffinch song. A better song at the learning stage does not, for instance, allow it to defend a territory or court females more effectively — although it may well have such a consequence later, after the song has been learnt (p. 127).

Further experiments have been carried out to discover what kinds of sounds a bird will learn to sing. Will it only learn the song of its own species or will it learn any tape recorded song it may have played to it during its memorizing phase? The results appear to differ among species. Chaffinches will only learn songs that are similar to the normal song of the species. But young male bullfinches, or zebra finches, are much more flexible. They learn to sing whatever song their parent, or foster parent sings. A bullfinch reared by a canary will sing a canary song; and the male offspring of the aberrant male bullfinch will, in their turn, learn a canary song too. Birds, such as parrots and mynah birds, that specialize in vocal mimicry, can learn to reproduce an even wider variety of sounds. Learning, then, is important in the development of all song birds, but the range of what different species learn differs considerably.

3.5.4 Memory

Memory is, by definition, essential for learning; and memory has, like learning, been studied more in the laboratory, with respect to artificial behaviour patterns, than natural behaviour. It is, however, a part of many described feats of animal behaviour, as we may illustrate with the Bearends's study of the digger wasp *Ammophila*, (Figure 3.11).

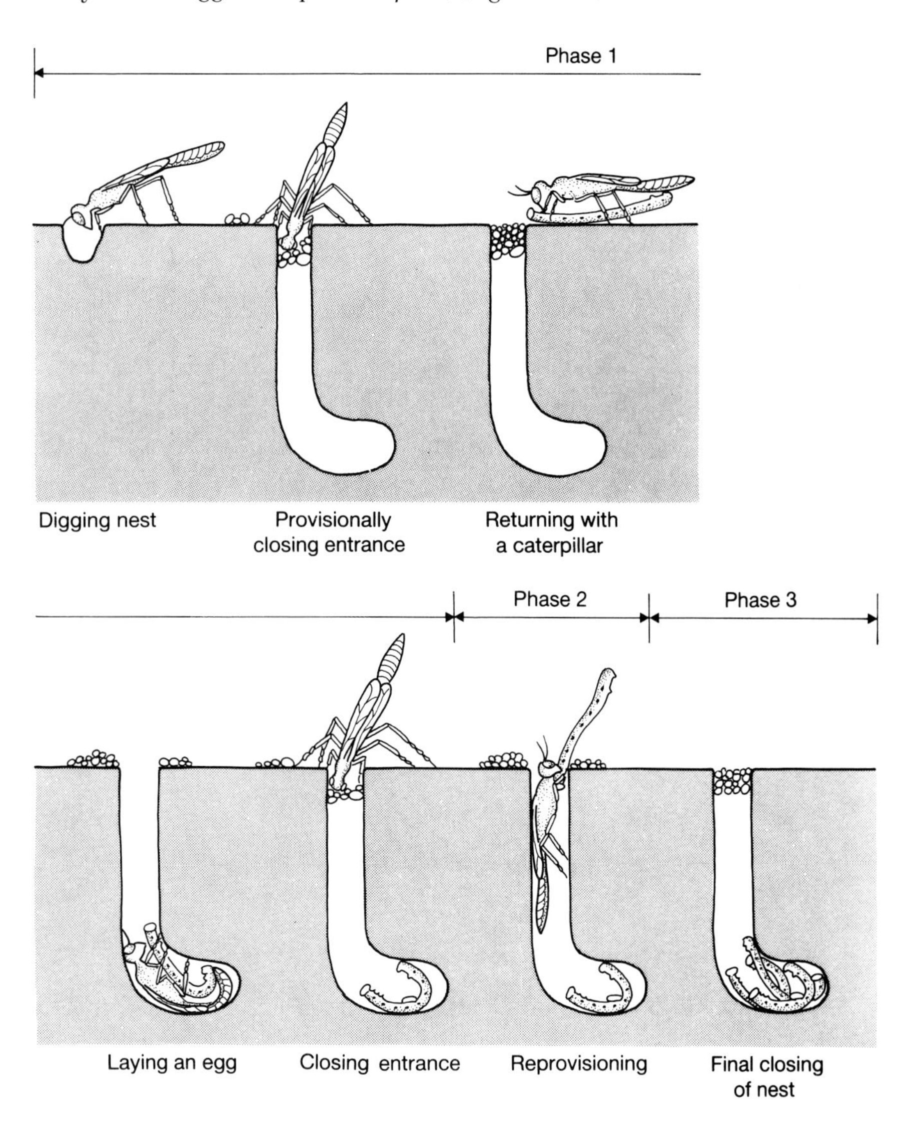

Figure 3.11 The main events at the nest of a female digger wasp. (After Tinbergen)

Ammophila campestris females dig burrows in sand. After a female has dug a burrow, she closes the entrance, and flies off to catch a caterpillar. She brings the dead caterpillar back to the burrow, and lays an egg on it. She then catches further caterpillars, one at a time, before finally closing the nest and leaving her offspring to develop by itself, feeding on her provisions. Such is the cycle of any one burrow; but, the Baerends discovered, a female may work on two or three burrows at a time. Each burrow may be at a different stage. The wasp will remember exactly what to do at each burrow, according to its stage in the cycle, and the number of caterpillars it already contains, even though she may not have visited it for several days. That would not be possible without a memory faculty, which can remember times, locations, numbers, and other information.

How do animals remember? What is the mechanism, or mechanisms? We can get some idea from certain experiments that have been done with rats. The experimenter first teaches his rat to perform some trick of other, such as running fast across its chamber. If the rat does not run when given some signal (such as a light coming on), the experimenter gives it an electric shock. The rat soon learns to run when signalled. The experimenter then waits for a certain amount of time and repeats the experiment to see how well the rat has remembered what it has to do. The interesting result is the relation between how well the rat remembers, and how long ago it was taught its trick. There is not, as we might have expected, a simple decline in how well it remembers as time passes. If the rat is tested just after being taught it does well. Rats tested up to about 12 hours later do less and less well. However, then the rat's memory seems to improve. A rat tested about 24 hours after being taught does as well as one tested just after being taught, and much better than one tested 12 hours after being taught. The experiment has been extended by trying to teach rats, at various intervals after teaching them to run to avoid a shock, to stay dead still to avoid being given a shock. One rat might be taught to run, and then taught three hours later to stand still; another rat might be taught to stay still about six hours after it was taught to run. In these experiments, the rats learn the new trick best at those times when they remember the old trick least well: it is as if a rat can learn a new trick more easily when its memory is not muddling it with the memory of the old trick.

There are two main kinds of theory of why animals forget: the decay theory and the interference theory. According to the decay theory, the memory of some event fades with time unless continually upgraded. According to the interference theory, animals forget things not because memory fades but because other memories displace them. The memory of more recent things interferes with the recall of things memorized longer ago. The rat experiments that we have been considering suggest that the interference theory is more accurate. The rat finds it easiest to memorize (learn) a new trick when the memory of the old trick is weakest.

In summary, then, animals undoubtedly can modify their behaviour as a consequence of their experiences. The exact nature of the influence of experience has been studied in many systems. It is undoubtedly important in the development of behaviour, and makes sense from the point of view of fitting the animal's behaviour to its environment (that is, learning is generally adaptive in nature). But the most striking feature of the research on behavioural development is *not* the steady accumulation of knowledge of particular influences of experience and other factors in development. The most discussed subject has instead been an abstract, conceptual controversy. Indeed, of all the parts of ethology, development has been the most subject to conceptual confusion, and has been the centre of the greatest controversy. Now that our account of the concepts and positive knowledge of development is complete, we can most easily understand that controversy.

3.6 The instinct controversy

The word instinct is now unfashionable among the scientists who study animal behaviour; but it has not always been so. Until about 1950 it was a normal part of the vocabulary. Then, in the 1950s and 1960s, it became highly controversial; it has now gone the way of all controversial terms — it is too highly charged to be useful. Although the controversy itself is now past history, it is worth knowing about, for the points of principle are highly important.

The main figure in the story is Konrad Lorenz, who began his work on animal behaviour in about 1930. It was then the heyday of 'stimulus-response' theories of animal behaviour, according to which all behaviour patterns are learned responses to associated stimuli. Lorenz's own observations led him to take a different view. He was struck by how similar the behaviour of different species can be, as in (for example) the courtship of different species of ducks which grow up in very different environments. It seemed impossible to explain the similarities by similar learning experiences. He duly explained them instead by inheritance, independent of the environment. He called behaviour that (he believed) develops independently of experience instinctive. Of course he did not deny that experience does influence the development of some behaviour patterns, and he accordingly divided behaviour patterns into instinctive (inherited patterns which develop independently of experience) and learned (the opposite). He also suggested a method to distinguish to which category any given behaviour pattern belonged; this was the isolation experiment. We have met isolation experiments before (p. 59), but allowed them then only a narrower purpose. In an isolation experiment, animals are isolated from their normal environment at birth; they have no opportunity of normal learning. According to Lorenz, if a behaviour pattern developed normally under such circumstances, it belonged in the instinctive, rather than the learned, category of behaviour.

Lorenz's argument might seem to move from a justified limited conclusion to an unjustified general one. For an isolation experiment cannot eliminate all possible environmental influences, all sources of experience; it cannot prove a universal negative, if only because universal negatives cannot be proved. Strictly speaking, it cannot be shown that a behaviour pattern develops independently of experience; only specific, identified factors may be ruled out. Indeed, in a sense, behaviour *cannot* develop independently of the environment, and cannot, in a developmental sense be called 'inherited' at all. Behaviour is part of the phenotype of an organism, not its genotype; behaviour develops by an interaction of genes and environment. An account of the development of behaviour within an individual would have to mention a series of environmental influences on gene expression. One cannot refer to a behaviour pattern as inherited (or instinctive) or learned; the terms can only properly be used to refer to the causes of *differences* between individuals. If one hive of bees catches foul brood, but another does not, we can legitimately ask whether the difference is genetic or environmental in origin, but that is quite different from asking whether the behaviour itself, in its development, is genetic or environmental. It refers only to the cause of a difference not to a whole course of development.

To identify the cause of a difference as genetic or environmental in origin, although it is a clear dichotomy, does not save the original Lorenzian dichotomy. For whether a difference is genetic is independent of whether the different types of animals concerned are or are not influenced by experience during development. It is logically possible that the difference between the behaviour of two types of animal could be genetically caused, but that both types could learn the behaviour in question. This would be a case of a genetic influence on learning, of which several real examples are known. Some genetically different strains of rats, for example, differ in their ability to learn to run through mazes. Likewise, a difference may be environmentally caused, but the two types not be learnt. There are two types of locust, for example, the solitary and migratory types, that are so different in form and behaviour that they were once mistakenly classified as different species but the difference is environmentally triggered (by the degree of crowding) and the two types are not genetically different.

All these points have been repeatedly raised against the original Lorenzian dichotomy. Daniel Lehrman is perhaps the best known critic, but there were (and are) many more. Lorenz responded by saying that he did not hold the position they were attacking with such gusto. On the first page of his book *Evolution and Modification of Behaviour* (1965) he wrote 'no biologist in his right senses will forget that the blueprint contained in the genome requires innumerable environmental factors in order to be realized in the phenogeny (development) of structures and functions'. He went on to clarify his distinction. In dividing behaviour into inherited and learned, he was not so much trying to explain how behaviour develops as how it comes to be adapted. He agreed that behaviour is not inherited, but what he called the 'adaptive information' of behaviour is inherited. By 'information' Lorenz means that, in order for an animal's behaviour to be adapted to the environment, the animal must have some knowledge (information) of the environment: if an animal is to perform some adaptive behaviour, such as finding food, it must know what food looks like.

Now, according to Lorenz, the dichotomous distinction between instinctive and learned behaviour cannot be made for development, but it can for adaptation. The adaptation of any piece of behaviour to the environment could have either of two sources. It could be due to natural selection over the generations; or it could be learnt. A bird pecking at food grains could have known without learning what food looks like, or it could have learnt it. In this case, as we have seen, there is a bit of both, but the dichotomy is not logically false as it was when applied to development. 'Adaptive information' can be said to be inherited, if natural selection has built that information into the animal's genes over the generations.

Whether Lorenz's critics were firing a straw man is unimportant. Only the conceptual conclusions really matter. Instinct has become unfashionable because of its association with an erroneous theory of development. Behaviour cannot be developmentally divided into the inherited and the learned (or environmental); all behaviour is influenced by both factors. Only differences between classes of individuals can in principle be attributed to a single cause. Lorenz's point, however, that the adaptiveness of behaviour is due to inherited information, remains both valid and important.

3.7 Culture in animals

The distinctive property of cultural behaviour, as ethologists use the term, is the way it is passed on from one generation to the next. Instead of being inherited by the process of Mendelian genetics, it is 'inherited' by imitation. An animal acquires the behaviour pattern by imitating it from another. Therefore, all that is necessary for a species to acquire a culture is that its members should be capable of learning and memory, and meet other members of their own species sufficiently often to be able to learn things from them. Cultural behaviour is therefore most likely to be found in species that form social groups.

The clearest examples of cultural behaviour do indeed come from a social animal, the Japanese macaque, a monkey that inhabits the forests of various islands around Japan. Since 1950 Japanese ethologists have been watching troops of the Japanese macaque and recording the lives of individuals within the troops. In 1952 they started leaving sweet potatoes on the beach at the forest edge, as food for the macaques, which duly came out of the forest and ate them. Next year something new was seen. Macaques were picking up potatoes, taking them to the sea, and washing the sand off the potatoes; they used one hand to dip the potato into the sea, and the other to brush the sand off. Potato washing was an entirely new pattern of behaviour; no macaque had ever been seen washing a potato before. The habit was invented by a single two-year-old female called 'Imo', who, in a moment of inspiration, had thought of washing the potatoes. Soon other macaques in her troop imitated her and the habit became more widespread. The first macaques to copy Imo were those of her own age; older macaques learned the trick later. Thus, five years after Imo's invention, 80% of the macaques in the troop between the ages of 2 and 7 washed potatoes; but only 18% of those eight years or more washed potatoes, and all 18% were females. There are two reasons why the habit spread faster among the younger macaques. One is that they are more willing to explore new skills; the other is that Imo herself was young, and macaques interact most with other individuals in their troop of a similar age to themselves.

Imo's career as an inventor did not end with potato washing. Two years later she surpassed this. As well as leaving potatoes, the ethologists had also scattered grains of wheat on the beach. The macaques initially picked out single grains, one at a time, from the sand; but Imo learned to pick up whole handfuls of sand and grain mixed up, and throw the whole lot into the sea. The sand sunk; the wheat floated: and Imo could then skim a clean meal of wheat off the surface. Wheat skimming is rather cleverer than potato washing. Washing a potato is only a small development from brushing earth off it, which is something all macaques do anyway; but the separation of sand from wheat required Imo to throw away the food after she had picked it up, and then wait for the sand to sink, before she collected her food up again. Once learned, it is a much better way of collecting wheat. Like potato washing, it was soon imitated by other macaques in the troop. The first to learn were again those of a similar age to Imo. From them, it passed into the culture of the troop.

The habit of opening milk bottle tops spread through several species of birds by an analogous cultural process. The habit is best known in tits, particularly great tits (*Parus major*) and blue tits (*P. caeruleus*); but it is found in other species too. It was first described in England, near Southampton in 1921, birds were seen removing the tops of milk bottles and drinking the milk beneath. Through the 1930s and 1940s the habit quickly spread through Britain, at far too fast a rate for it to have taken place by the natural selection of Mendelian genes (Figure 3.12). J. Fisher and R. A. Hinde, who first collected the reports of milk bottle opening by birds, therefore suggested that the habit might have spread by imitation. After a tit had seen another tit open a milk bottle top it might then try out the behaviour pattern for itself; it would discover the reward, and go on to open other bottles. So the habit would spread.

Imitation need not have been the only process at work. From the maps of

Figure 3.12 The spread of the habit, in tits, of opening milk bottle tops in Britain from the first record in 1921 near Southampton until 1947. (After Fisher and Hinde 1949, photos by V. L. Breeze)

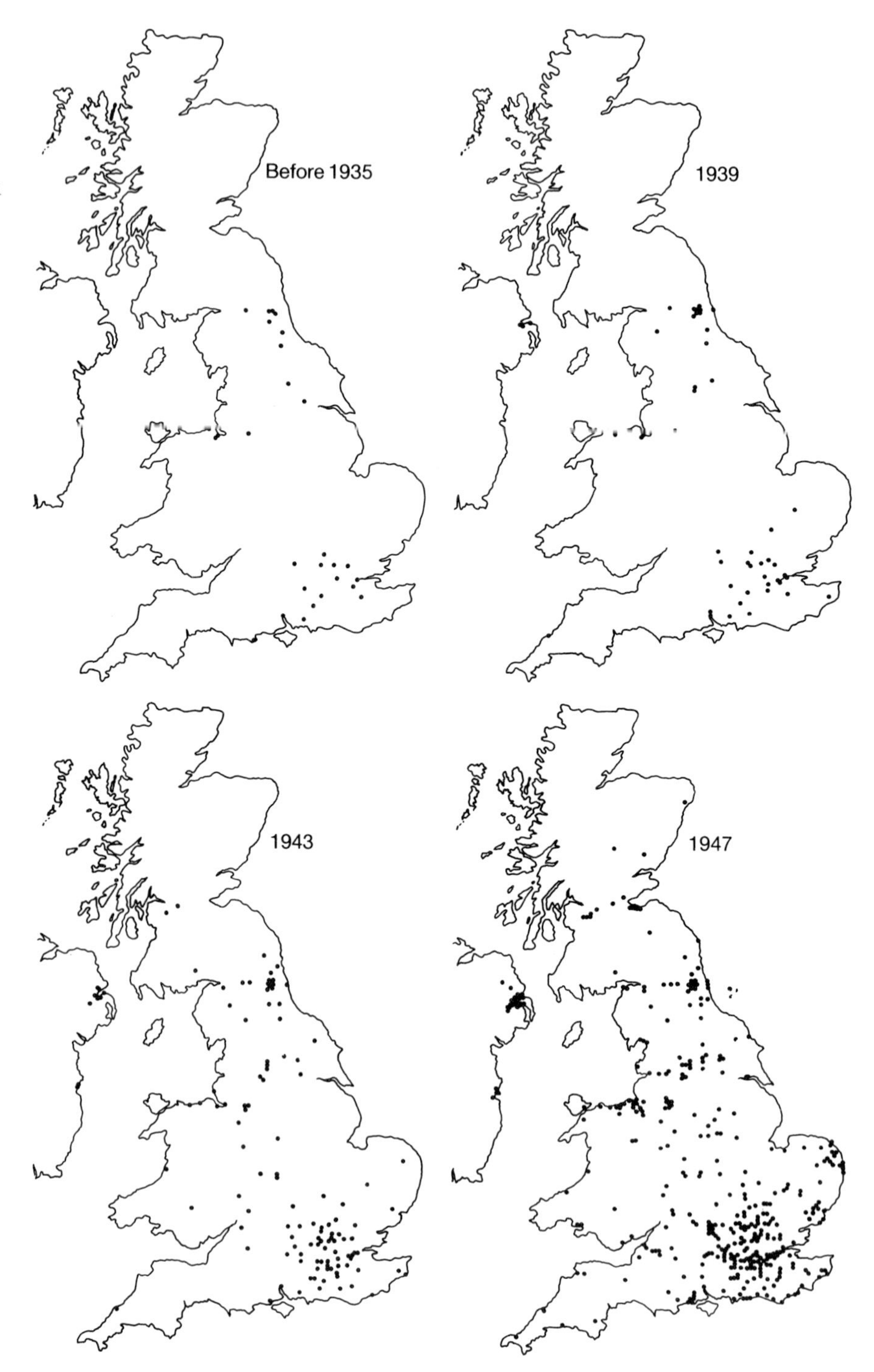

the spread of the behaviour (Figure 3.12) it looks as if the habit was invented independently more than once. A recent experiment, however, confirms that birds can learn the habit by imitation, and also adds another side to the story. D. F. Sherry and B. G. Galef experimented on the black-backed chickadees (*Parus atricapillus*) of Canada (these are close relatives of the British tits in which the spread of the habit was recorded). Their experiment was as follows. They first presented small milk containers 'of the type often provided with coffee in restaurants', with intact tops, to 16 birds. Of the 16, four opened the containers spontaneously, the other 12 did not. These 12 made the main experimental subjects.

They divided the 12 into three groups of four birds each. The birds of one group were put, each with an unopened milk container, in one compartment of a cage that was divided into two compartments by a wire mesh through which the birds could see. In the other compartment Sherry and Galef put a 'tutor' bird — one of the four that already knew how to open the milk containers — and a milk container. The bird in this treatment would have the opportunity to learn the skill by imitation. The birds of a second group were put alone in a cage with a full milk container that had already been opened by the experimenters. They were treated in this way because birds might be able to acquire the skill simply by coming across an already opened milk bottle. On seeing it, the bird might drink out of it. The bird might then learn that a certain movement of its bill in relation to a milk bottle results in a meal; and if the bird was then to perform the same activity on an unopened bottle it might break through the top for itself. In this case, the spread of the habit would still be by a non-genetic learning process, but not by imitation. Sherry and Galef's third treatment was a control: the birds were put alone, each with an unopened milk container, as in the first part of the experiment. In the final stage, the 12 birds from all three treatments were re-tested for their ability (when alone) to open one of the milk containers. The result (Table 3.2) was that three of the four tutored birds had learnt to open the milk containers, as had three of the four that were given an already opened one; but none of the controls had learnt the skill.

Sherry and Galef's experiment supports the original hypothesis, that the spread of milk bottle opening was partially driven culturally, by imitation. But it suggests that this was not the only process. Some birds probably learned the skill after encountering already opened bottles, from which not all the accessible milk has been drunk by the original opener. The relative

Table 3.2 Sherry and Galef's experiment on the learning of black-backed chickadees to open milk containers. The text explains the three stages of the experiment. The three birds in the 'opened' treatment that drunk the milk in the training session did not have to open the container. The one bird in the control treatment that opened its container in the training session did not drink the milk. The difference between experimental and control treatments is statistically significant. (Slightly simplified from Sherry and Galef (1984))

		Number of birds opening milk containers in:		
	N	Pre-training	Training	Testing
Spontaneous opening	16	4	—	—
Treatment				
Tutored	4	0	3	3
Opened	4	0	3	3
Control	4	0	1	0

importance of the three processes by which the habit could have spread (spontaneous discovery, imitation, and learning from already opened bottles) is more difficult to assess. But the purely cultural process did operate in the experiment, and therefore probably occurred in nature as well.

Cultural inheritance is analogous to genetical inheritance and leads to a process of evolutionary change analogous to genetical evolution. Darwinian, genetical evolution takes place because genes are passed on from one generation to the next, and if some genes build better bodies than others, they are favoured by natural selection, become commoner, and evolutionary change will take place. With cultural behaviour, an analogous process will operate. If one behaviour is more readily imitated than another, it will become commoner in the population, and evolutionary change will take place. The two processes are only analogous, not identical, and there are important differences between them. One is their relative rates. Genetical evolution is necessarily slow, because genes are only passed on once per generation: gene frequencies can therefore only change once per generation. Cultural change can be much faster. Learning by imitation can take place in a few minutes, whereas the reproduction of a whole new adult animal can take many years. There is time for many cultural evolutionary events in each generation. Genetical evolution must therefore be a slower process than cultural change. The relative rates of the two processes have an important consequence for our understanding of humans. In that most cultivated of species, cultural changes proceed at an exceptionally high rate, and affect nearly all components of behaviour. We therefore might expect cultural changes to have greatly outstripped our slowly changing evolutionary heritage, in which case our present behaviour should be attributed more to the process of cultural, than of genetical, evolution. This is the reason why many ethologists are hesitant to apply the insights of the theory of natural selection, which have been gained for the social behaviour of animals with only rudimentary cultures, to the behaviour of our own species. Our culture may have rendered natural selection relatively irrelevant.

3.8 Summary

The inheritance of behaviour is effected by means of arrangements of molecules, called genes. The simplest scheme of inheritance is that for one pair of genes, controlling one pair of traits; the traits will then segregate in a 3:1 ratio in the second generation. One behavioural example that approaches this degree of elementary simplicity concerns hygienic behaviour in honey-bees. When inheritance is controlled by many genes, the cruder techniques of quantitative genetics can be applied. Artificial selection, for instance, has demonstrated a genetic influence on the mating preferences of ladybirds.

It is possible to discover what factors are *not* needed for the normal development of a behaviour pattern by isolating young animals from hypothetically necessary factors. If normal development takes place in the absence of the factor, it is not necessary. It has thus been shown that both practice and 'maturation' influence the accuracy of pecking in chicks. It can also be asked whether the influence of a factor is confined to a particular 'sensitive period'. Imprinting usually takes place in the first few days after birth.

Learning may be non-associative or associative. Habituation and sensitization, the two forms of non-associative learning, are shown by *Aplysia*, and the neurophysiological control is understood: habituation is caused by decreased release of neurotransmitter from the sensory neuron; sensitization by release of a chemical from an interneuron which renders the motor neuron more easily depolarized. Associative learning, often called conditioning, has been studied in two forms, illustrated by Pavlov's dogs, and pigeons learning to peck at discs in order to be fed. The learning of song in chaffinchs takes place in two phases, a memory phase and a practice one. In the memory phase it must be able to hear a normal song; and in the practice phase it must be able to hear its own initially crude efforts, which it compares with a 'template' of the memorized perfect song. Memory is needed in many natural situations, but has been most studied psychologically. Experimental results are most easily explained by 'interference' theories, in which the memory of one task is most efficient when no other task is being memorized.

Behaviour patterns cannot be developmentally divided into inherited and learned behaviour, for the development of every behaviour pattern is influenced by both factors. Behaviour itself is not inherited. The 'adaptive information' manifested in behaviour can, however, be described as inherited, because it is due to the action of natural selection over many generations.

A species capable of learning can develop a culture, consisting of behaviour patterns passed among individuals by imitation. Japanese macaques have in this way acquired a cultural repertory of feeding techniques, potato washing and wheat skimming. Tits have likewise, in part at least, learned to open milk bottle tops by imitating other birds. Cultures can evolve in a manner analogous to genetical evolution. Cultural evolution will probably be much the faster process of the two.

3.9 Further reading

Volume 3 of the series edited by Halliday and Slater (1983) covers much of the material of this chapter more thoroughly. Ehrman and Parsons (1982) introduce behavioural genetics. There is a recent conference on learning, (Marler and Terrace 1984). Kandel (1979), on *Aplysia*, and Dickinson (1980) and Staddon (1983) cover the subject more generally. Hailman (1969) describes his work on gulls. Bonner (1980) reviews animal culture.

4 / Movements and migration

4.1 The principles of migration

We shall use the term migration here to refer to almost any pattern of movements by living things. The mass movement of herds of wildebeest across the plains of Africa, the seasonal migrations of swallows and of red admiral and monarch butterfies, the return of salmon to their natal stream. All invite two questions: why do animals migrate and how do they know the way? Taking the first question first, an answer can at least be given in abstract terms. Each kind of animal lives best in a particular environment. There must be enough of the right food, the temperature and habitat must be right, there must not be too many other animals that would parasitize it, or kill it for food. In some cases, the environment is so constant, or the animal can live in such a wide range of environments, that once an animal has found a suitable place to live in it will not need to move far to satisfy all its bodily wants. Snails of the species *Cepaea nemoralis*, for example, usually die less than a hundred yards away from where they hatched. However, environments are usually so variable that even if the conditions are good where the animal is now, they probably will not be in a month's time; it may then be better to be a hundred miles south. The abstract reason why species migrate is that environments change so that the best place to be varies with time. The pattern of animal movements should follow the pattern of environmental change. For instance, if environmental changes are capricious, the animal's migration viewed in isolation will also be capricious.

The answers to the second question, of how animals find their way when migrating, can be more various. If the animal has a sufficient sensory range, it need only move towards areas that its sense organs reveal to offer better conditions. This kind of 'planned migration' is performed for example by wildebeest (Figure 4.1). Wildebeest (discussed on p. 102 below) inhabit the plains of Africa, and are frequently on the move. But they do not move blindly, in the hope of coming to better pastures. Wildebeest eat grass, and grass grows after rain; they can sense where rain is falling by using their eyes and ears (but not their noses): they then migrate in that direction. Wildebeest movements follow, at a distance of a few days, the pattern of rainfall; but they only keep moving so long as rain is falling or has recently fallen within the area scanned by their senses.

Planned migration is well suited to the capricious pattern of local rainfall in East Africa. But other environmental changes are more predictable. The

Figure 4.1 Great herds of wildebeest migrate around the Serengeti Plains of Kenya, East Africa. Their migrations are directed towards areas of recent rainfall. (photos: Heather Angel)

animal can anticipate them, rather than waiting for direct evidence. In Temperate regions there are regular seasonal cycles, ultimately driven by the regular cycling of the Earth around the sun. Associated with the cycle of day-length and temperature are many other cyclic changes that matter to many animals: cycles of plant abundance, of leaves on trees, of the insects that live on plants. It is not surprising, therefore, that many animals perform regular seasonal migrations, northwards in the spring and southwards in the autumn (in the northern hemisphere). The insects (mainly Lepidoptera) and birds that do so cannot directly sense the superior environment of the north (in the spring), but its superiority, on average, is guaranteed by the predictable changes of the seasons.

The seasonal migrator needs an accurate cue to anticipate the seasons, and at least a compass sense to guide it. They measure the seasons by the changing day length, as may be demonstrated by keeping migratory birds under artificial day length conditions: they then migrate at a time dictated by their experienced day length, rather than the time of year. The effect of day length on behaviour is mediated hormonally; the birds are prepared to mi grate in the autumn by declining production of sex hormones at the end of the breeding season; they do not migrate if injected with sex hormones. Similarly, the release of sex hormones in the spring, which stimulates reproduction (p. 46), also stimulates northward migration, an association that led the biologist J. B. S. Haldane to remark that although 'we must be very careful in attributing human motives to animals, the emotion behind migration to breeding places is almost certainly more like human love than hunger or curiosity.' Seasonal migrators could in principle orient themselves by a simple sense of direction; all they need to do is obey some such rule as 'fly two hundred miles south' or a more complicated series of directions. They could obtain compass information from the sun and stars, and it is indeed known that some seasonal migrators follow stellar patterns; stellar orientation has been particularly well studied in the indigo bunting, a bird that inhabits North America. Probably most such birds possess more powerful navigational skills, such as those we shall discuss shortly for pigeons, but it takes special experiments to demonstrate the fact.

Most insects are not powerful enough fliers to be able to carry out seasonal return migrations; but a few can. The monarch butterfly (*Danaus plexippus*) is one. It is a large insect, strikingly coloured in black and gold, which lives mainly in North America. In the summer the monarch is distributed from Mexico to Canada, but with falling temperatures it moves south, and in the winter it is only found in Mexico and the southernmost United States (Figure 4.2). In the spring they migrate north again. Not all monarchs migrate south in the autumn; some hibernate in the north. Those that do migrate move at astonishingly high speeds, of up to 100 miles a day. If the winter is warm in Mexico they live as free individuals as in the summer, but when it turns cold they aggregate in dense groups, and remain still, in order to con-

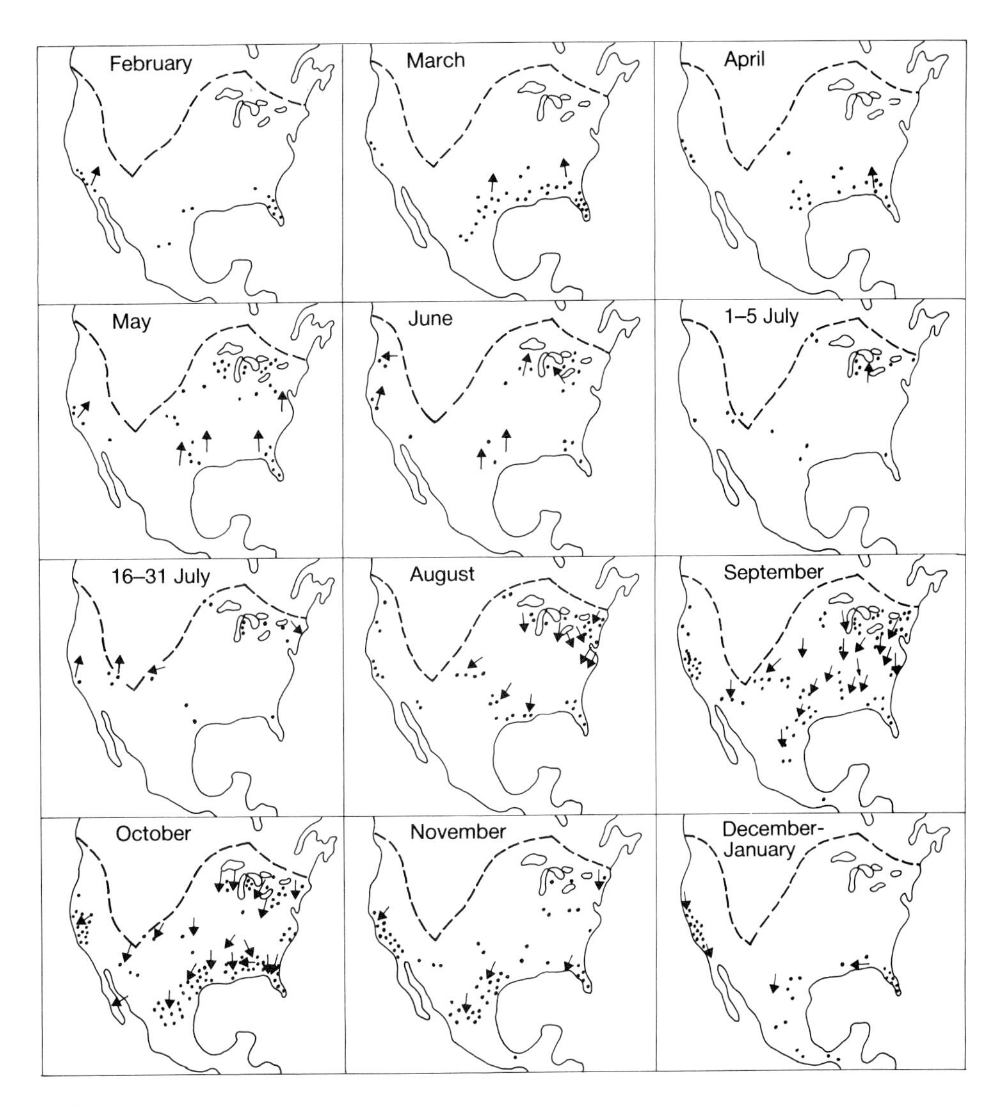

Figure 4.2 Range maps of monarch butterflies through the year. In the autumn they migrate from the United States and Canada southwards towards Mexico. In the spring they fly North again. It is an example of a 'seasonal return migration'. (After Baker)

serve energy. For a cold blooded insect, the main reason for seasonal migration is probably the need to be in a warm enough place to allow an active life; a warm blooded bird does not have that problem.

We have considered why animals migrate; we might also remark on its ecological consequences. Migration can regulate the local density of animals. If the density of animals in an area increases to the point of overcrowding, there will be insufficient food to go round, and it might pay an animal to move away. If competition is too great in one place, it may be better to try another. For this reason, an increase in population density often precipitates a round of emigration. The migrations of lemmings are stimulated by overcrowding; so too are the movements of aphids (insects that live on the sap of plants including greenfly). Most aphids can grow up either as a wingless stationary form or a winged migratory form; they are more likely to grow up with wings if the local population density is high. The regulation of population density can only be a consequence of migration, not the reason why natural selection causes the habit to evolve. Natural selection only favours habits that make organisms leave more offspring; the advantage of a habit must therefore be in the short-term. Population regulation, however, if it has any advantage at all, can only be a long-term one; it must therefore be a consequence of individual decisions to emigrate, taken on the grounds that conditions will be better elsewhere, not on the grounds that the population level must be kept down in order for the local resources to be conserved. Natural selection takes no account of long-term considerations.

4.2 Homing

4.2.1 Local landmarks and home cues

The life-cycles of many animals require them to find their way back to a particular place. The task for a green turtle (Figure 4.3) to find a tiny island in the expanse of the Atlantic Ocean, or a salmon to find the exact river tributary in which it was born appear to us exceedingly difficult. Because the purpose of homing is often self-evident (it might, for instance, be a matter of finding the

Figure 4.3 Green turtles (*Chelonia*) live most of their lives in the sea, but return to their natal island to breed. They deposit their eggs in the sand on the beach. Different populations probably return to different islands; the most famous such island is Ascension Island, in the Atlantic Ocean. This green turtle female has homed to Aldabra Island in the Indian Ocean. (photo: Tim Guilford)

right place to lay eggs), the question of how they find their way is usually more interesting. We might distinguish three possible answers. One is that animals memorize local landmarks and directions on their way out, and simply reverse the directions to find their way home; a second is that the home site itself has some property that can be recognized at a distance. The third, which is most likely to be important in long-distance homing, is that the animal has an internal 'map' sense, and can both estimate its own map reference, and knows that of the home site. We can consider reasonably clear cut examples of the use of local landmarks and of home stimuli, but when we come on to a possible map sense we shall move into one of the more unsettled areas of the science.

Let us consider first the homing problem of the bee-killing digger wasp *Philanthus triangulum*. Female wasps of this species dig burrows in the sand, to provide a nursery for their offspring. Each female digs several (about 6 or 7) cells off the side of her burrow, and lays an egg in each cell. When the egg hatches into a larva it will need food. That is where the bee-killing part of the wasp's name comes in. The mother wasp goes out of her burrow, catches and kills a bee (by stinging), and brings it back to the burrow. She then opens the entrance to the burrow and takes the bee down to one of the cells. She continues to catch and bring back bees, one at a time, until each larva has about two bees to eat. The mother wasp, therefore, does not merely dig a burrow, and later leave it never to return: she departs from and comes back to it many times. On every return she has to find her burrow, distinguishing it from its surroundings. It is not an easy task to find a particular burrow, because these wasps can nest in quite dense groups; there might be more than twenty burrows within a circle of five yard radius. The problem of how the digger wasp locates her home has a special place in the history of ethology: it was one of the first questions about behaviour mechanisms ever to be asked, and experimentally answered. It was studied by Niko Tinbergen in 1929, on the heaths and sand dunes of Hulshort in Holland.

Tinbergen first confirmed, by individually marking all the nests and wasps in a particular area, that each wasp does indeed always return to her own burrow. He then performed some simple experiments to test whether the digger wasp recognized her own entrance by the distinctive array of odd objects (haphazardly fallen sticks, pine cones, stones etc.) around it, or by some stimulus emanating from the entrance itself. He placed around the entrance of each of several chosen burrows a neat circle of pine cones. He left them there for a few days, checking that the wasps still kept returning to their burrows. He then waited for the wasps to leave on bee-hunting expeditions, and while they were away he moved the cones a yard or so away from the entrances (Figure 4.4). When the wasps returned they landed where the entrances 'should' have been, in the centre of the circle of pine cones. Evidently they recognized the entrance by its surrounding landmarks, not by any stimulus from the entrance itself.

Learned local landmarks are a feasible means to navigation in a local, familiar area. But longer distance homing must require other techniques. Let us now consider the techniques of the salmon. All species of salmon have similar life cycles. They are laid as eggs in river tributaries all over the northern hemisphere. They live their first few months in the river, then migrate downstream to the sea; they spend two or three years in feeding and growing out at sea, during which they cover thousands of miles; and when mature, they migrate back to exactly the same river, and same tributary, as they were born in. How do they find their way home? The answer is thought to be, mainly by smell. The attractive odour may come from either (or both) of two sources: the young salmon which are still in the stream, not yet having migrated to the sea, and any other characteristic odours in the stream. Salmon have been shown to be capable of the necessary olfactory discrimination, but the most direct evidence that they use their sense of smell comes from experiments, of the kind first performed by W. J. Wisby and A. D. Hasler, in which the salmon's olfactory sense was impaired. Wisby and Hasler

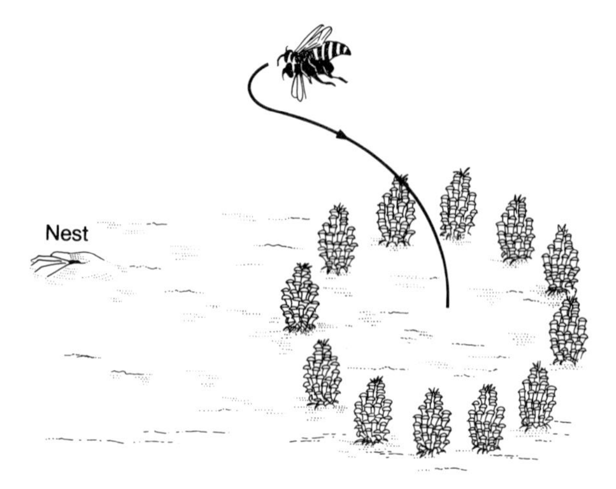

Figure 4.4 The bee-killing digger wasp *Philanthus* uses physical landmarks to recognize its home nest. In an experiment, Tinbergen allowed wasps to become accustomed to a circle of pine cones around its nest entrance; he then moved the circle while the wasp was away, and on its return the wasp sought the entrance of its nest in the circle of pine cones as before. (After Tinbergen)

Table 4.1 Numbers of coho salmon released and later recaptured in two rivers in Washington, with and without their olfactory sense impaired. Salmon with their noses plugged homed less accurately. For locations, see Figure 4.5 (From Harden Jones 1968, using data of Wisby and Hasler).

Stream of origin		No. released	No. recaptured	
			Issaquah	East Fork
Issaquah	Controls	121	46	0
	Nose plugged	145	39	12
East Fork	Controls	38	8	19
	Nose plugged	38	16	3

actually plugged the noses of their salmon, which then homed less accurately than untreated controls (Table 4.1); in more recent experiments, the same result has been obtained by cutting the salmon's olfactory nerves. Experiments in which young salmon have been transferred from their stream of birth, to be released in another stream, have been used as evidence that salmon imprint on the smell of their river during a sensitive period just before they migrate downstream. In one such experiment, L. R. Donaldson and G. E. Allen took 72,000 young salmon at the 'fingerling' stage (when they are about one year old) from the Soos Creek Hatchery in Washington (for locations see Figure 4.5) and divided them into two groups. In 1952 they released one group (identified by the removal of the right pelvic fin) at Issaquah Hatchery, and the other (which had their left pelvic fin removed) at the University Hatchery. The salmon returned in the winter of 1953/4; they homed on their release sites. None of them returned to Soos Creek, the site of their birth and first year of life. Of 71 marked salmon caught at Issaquah Creek, 70 lacked their right pelvic fins; all 124 marked salmon caught at the University Hatchery lacked their left pelvic fins. Results like that suggest that the young salmon learn the odour of their native stream (or of the young salmon in it), and later find their way home by seeking that scent. But there is another explanation. Salmon migrate in schools. The results of the transplantation experiment may only be due to the transplanted salmon following the lead of the native salmon. Recent commentaries, therefore, such as the review by O. B. Stabell, maintain that the case for imprinting is at best not proved. But it is not in doubt that salmon rely critically on their sense of smell to guide them home. Whether they learn their home stream's distinctive smell during a discrete sensitive period is undecided.

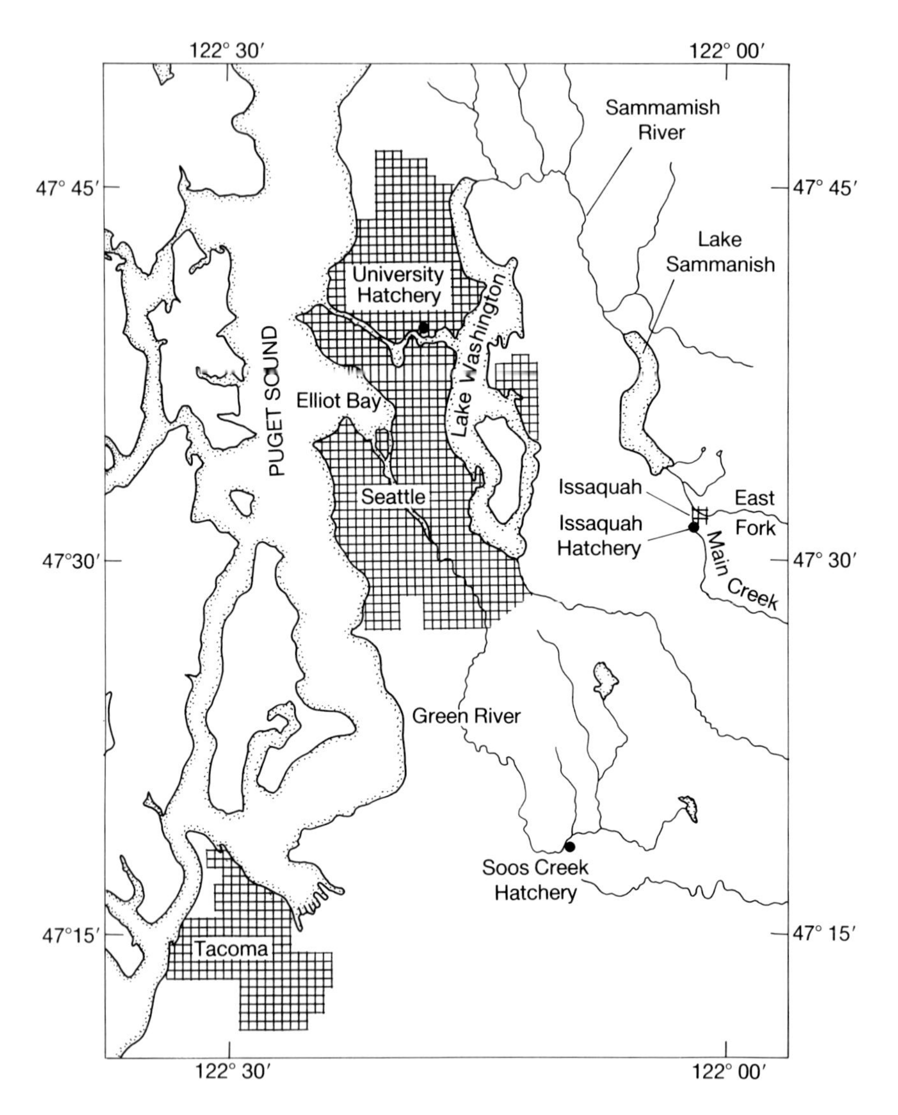

Figure 4.5 The locations of Issaquah Creek and the University Hatchery are shown on this map of part of Washington State, U.S.A. Experiments on the use of olfaction in homing by coho salmon were performed here. (After Harden-Jones)

4.2.2 Pigeon homing

It is less easy to believe that the large scale migrations of birds are accomplished by memorized landmarks and home cues than are those of the digger wasp and salmon. Moreover, experiments on pigeons have been thought to rule out that possibility. Pigeons taken from their home loft to some place, say one hundred miles to the east, which the pigeons have never visited before, and then released, are capable of finding their way home. After release the pigeon circles a few times around the release site for a few minutes and then flies off, usually in the approximate direction of its home loft. A few hours later it will arrive home. Then, a few days later, after the pigeons have had time to rest, they may be taken to some other point, perhaps one hundred miles south this time, and they will repeat the trick. Homing pigeons therefore appear to be able to find their way home from any starting point (provided it does not exceed their maximum flight distance), whether or not they have seen their starting point before. We shall call this ability 'true navigation': the ability to find the way through unfamiliar areas.

True navigation is not just a matter of finding the way home from a familiar starting point, which can be achieved (as by the digger wasp) with the use of memorized landmarks and learned orientations to them. Nor can true navigation be achieved only by 'compass orientation'. A compass indicates a direction and compass orientation means moving in a set direction relative to a compass; for example, the animal might always fly southwards, regardless of starting point it would need some internal 'compass' sense but nothing more. Compass orientation, we have seen, can account for the seasonal movements of birds and butterflies, which could find their way by just flying in a particular direction. They probably possess more sophisticated navigational equipment, but seasonal movements alone do not demand such skills. The test between compass orientation and true navigation is to move an animal experimentally away from its normal tracks. If it just flies by the compass, it will show an equivalent displacement from its goal; if it can navigate it will reach its goal despite the displacement. The experiment has been performed on European starlings. Starlings migrate during the autumn from the area around the Baltic Sea to their wintering places in southern England, northern France, and Holland. One year, some ethologists caught and marked (with rings around the birds' legs) a number of starlings as they were flying over Holland, put them in some aeroplanes, flew them to Switzerland, and released them. The juvenile starlings now behaved differently from the adults. The juveniles continued to fly south-west, and wintered in southern France and Spain. The adult starlings, however, flew north-west to the usual over-wintering grounds (Figure 4.6). Evidently the young starlings used only compass orientation whereas the adults used true navigation. The same kind of compass orientation is shown by adult mallard ducks in the phenomenon called 'nonsense' orientation. When mallards are taken away from

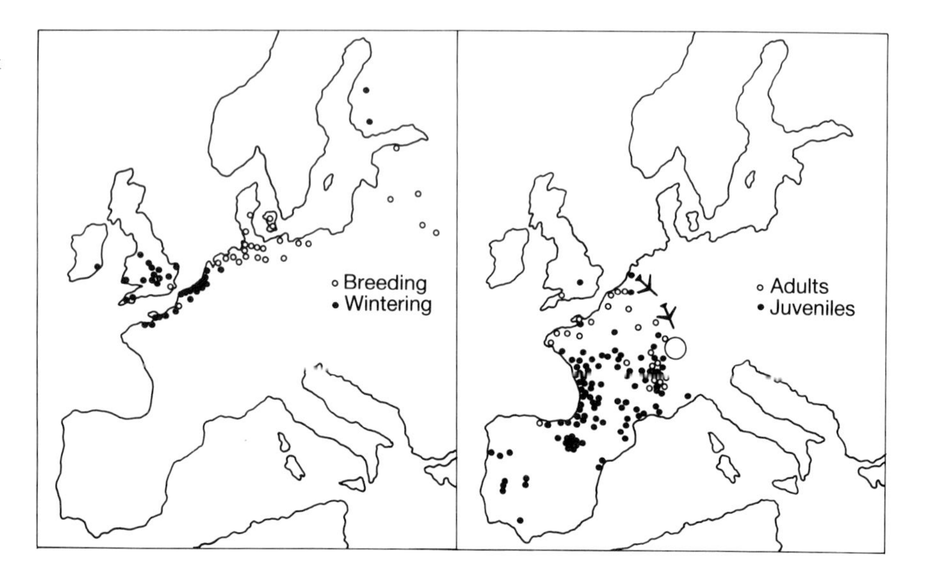

Figure 4.6 Starlings normally migrate south west during the autumn; the points on the left hand map show where starlings that were banded in The Hague, Holland, were later recovered in the winter and in the breeding season. On the right are the results of an experiment in which starlings caught over Holland in the autumn were transported to and released in Switzerland. Juveniles continued to fly south-west, but adults adjusted their direction to take them to their normal winter sites.

Slimbridge, in southern England, whatever the direction they are taken away they initially fly north-west, even if it is the opposite direction from home (Figure 4.7).

Most experiments, however, have been performed on pigeons, and as we consider the mechanism of true navigation we must concentrate on this one species. We should bear in mind continually that the whole field is controversial. Nothing is agreed, from fundamental data to the evidence for particular hypothetical mechanisms. Robin Baker has recently challenged whether the observations on pigeons require any true navigating skill at all. He suggests that the 'unfamiliar' release sites of pigeons may in fact be familiar, and the pigeons home using remembered landmarks and home cues (whether visual, auditory, olfactory, or magnetic). He claims that none of the release sites have been sufficiently far from the home loft to rule out the use of such cues, except in a few cases, such as trans-Atlantic displacement, and in these the evidence of homing is very weak. Apart from landmarks, pigeons could also home without any great navigational skill if they learned the direction on the way out. But this cannot be the only explanation, because pigeons can home normally even when taken out in enclosed vans in continuously rotating cages, or even under anaesthetic, both of which treatments should render learning the outward journey impossible.

No doubt experiments that unambiguously answer Baker's fundamental

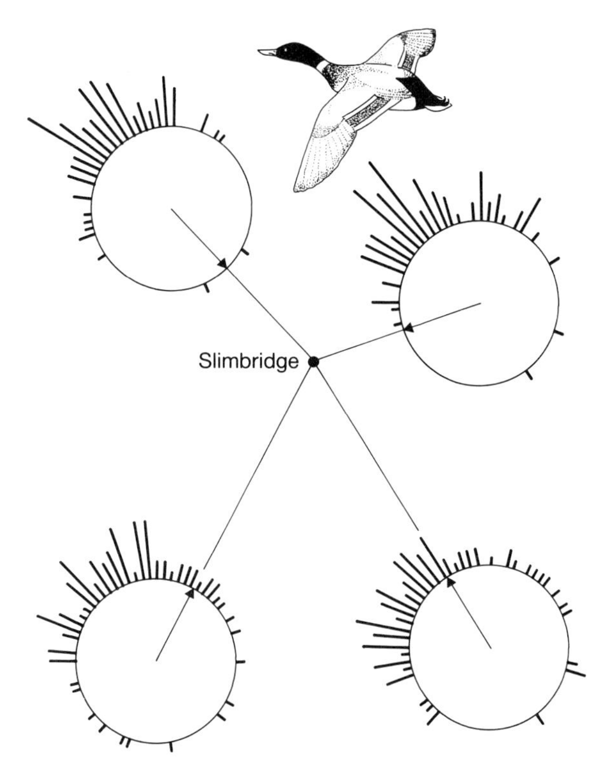

Figure 4.7 'Nonsense' orientation of a mallard duck. Whatever direction mallards are taken from Slimbridge, in England, they fly towards the north west after release. (After Matthews)

criticism will be performed, in which pigeons are taken to release sites which are clearly unfamiliar; but for now we must suspend judgement. It has been taken for granted by most workers in this field that pigeons can home from unfamiliar starting points. If we are to discuss their work we must share their assumption. Human beings seeking to navigate home from an unfamiliar site, would need a map and a compass. They would then remember map locations of home, identify their present map position, and then use their compass sense to find the direction from the one to the other. Experimenters on pigeons have accordingly searched for a 'map sense' and a 'compass sense' in these birds. The compass sense is the less controversial. Many experiments suggest that pigeons, as their first resort, use the position of the sun in the sky, making allowance for its daily changes by means of an internal clock. Thus, if a pigeon is trained on an artificial light/dark schedule, different from the outside, it will navigate with predictable error beneath the natural sun: pigeons, for instance, who think it is six hours later in the day than it is will

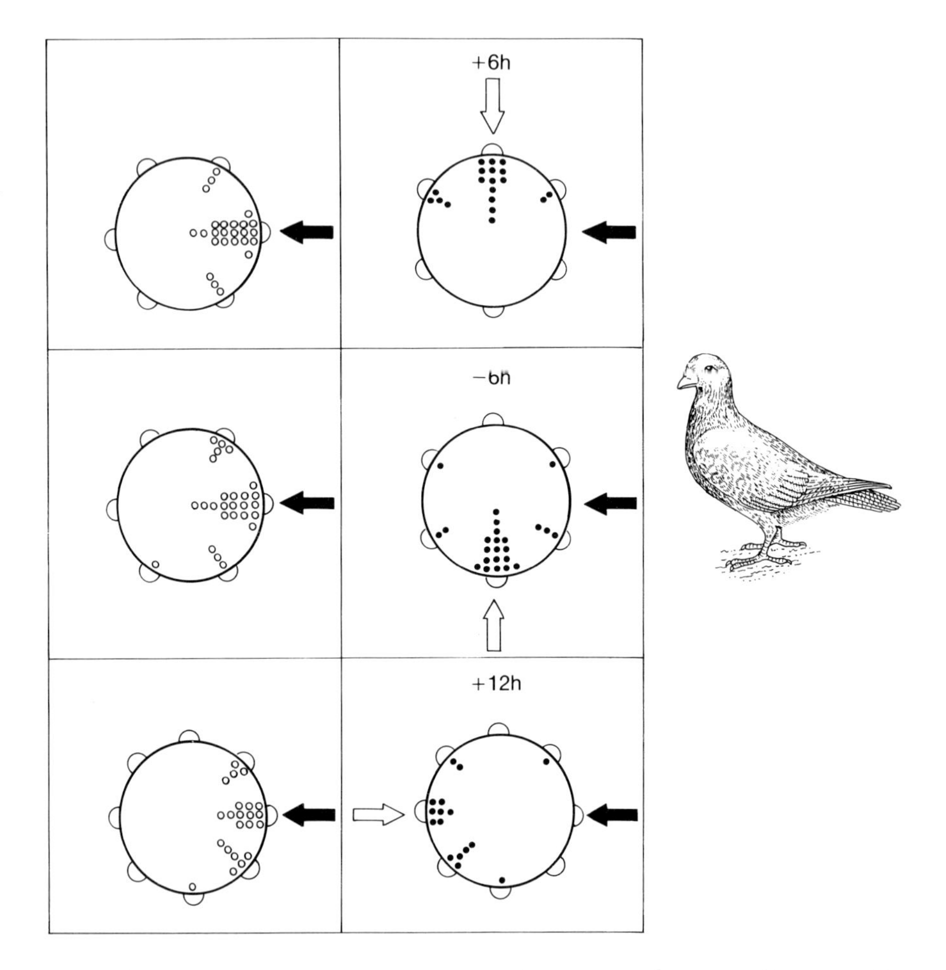

Figure 4.8 Clock shifts alter the orientation of homing pigeons. In this experiment, pigeons have been trained to peck towards a particular compass direction, which is that of the controls illustrated to the left. Each dot shows one peck by one pigeon. The directions of pecking by clock-shifted birds are shown on the right. The directions are those that would be expected if the pigeons use the sun as a compass, making allowance for the movements of the sun through the day. (After Baker)

head off 90° anti-clockwise away from the correct direction (Figure 4.8). Pigeons used to the northern hemisphere but moved to the southern hemisphere likewise orient with the predicted 180° error at noon. But pigeons must possess some other compass sense as well as the sun, because they can navigate correctly on overcast days, when the position of the sun is invisible. The experiments with clock-shifted pigeons, however, strongly suggest that the sun supplies their preferred compass.

The nature of any other compass sense is uncertain; there is, for instance, controversial evidence of a magnetic sense. But it is the map sense that is the real puzzle. Many imaginitive possibilities have been investigated, but we shall concentrate on two here — the sun and the earth's magnetic field — and a third possibility, that pigeons use olfactory home cues. The present verdict of many ethologists would be that the case for and against the olfactory and magnetic theories remains finally undecided; but the evidence does suggest that the 'sun arc' hypothesis is false. The subject is controversial, however. Let us take the 'sun arc' hypothesis first. We have seen that the position of the sun is used as a compass, by which pigeons direct themselves away from release points. It could also be used as a map. A map is most easily conceived as having two co-ordinates, like longitude and latitude on customary human maps. The sun arc hypothesis of E. V. T. Matthews suggests how the position of the sun could supply longitudinal and latitudinal positions. The pigeon would have to remember the position of the sun above its home loft at each time of the day. Then, at a release site, it could estimate the latitude by the height of the sun in the sky (through an estimated arc) and longitude by the time of day indicated by the sun at the release site relative to the time of day at home indicated by the pigeon's internal clock. If its home clock said the time was 12.00 a.m. but the sun at the release site indicated 6.00 a.m. the pigeon would infer it had been moved one quarter of the way round the world to the west. The hypothesis can therefore be tested by clock-shifting experiments. Pigeons can be trained to think the time is different from that outside their home loft, and then observed to see if they misorient as the sun arc hypothesis predicts. Pigeons clock-shifted 6 hours forward should behave as if they had been taken west: a pigeon, for example, clock-shifted 6 hours in a loft at Rome and then released at midday would think that the time at home was 6.00 p.m., deduce that it had been moved to New York, and therefore home by flying east (if we ignore the fact that it would recognize local landmarks). Even if displaced through three time zones eastwards, it would still infer it was over the Atlantic, and home in the wrong direction. But when the appropriate experiment was performed, the hypothesis proved wrong. The pigeon's compass sense only was disrupted, as in the previous clock-shifting experiments we discussed. A pigeon clock-shifted 6 hours early but taken through three time zones to the east would in fact behave as if it 'knew' it should home to the west, but because its compass bearings are rotated 90° clockwise (it thinks it is midday when it is 6.00 p.m.), it would 'home' towards the north rather than the west. The relevant experiment, by Walcott and Michener, has not escaped criticism; but it is not the only piece of evidence to go against the sun arc hypothesis, and even if that hypothesis has not been definitely refuted, we can say that such imperfect evidence as there is counts against it.

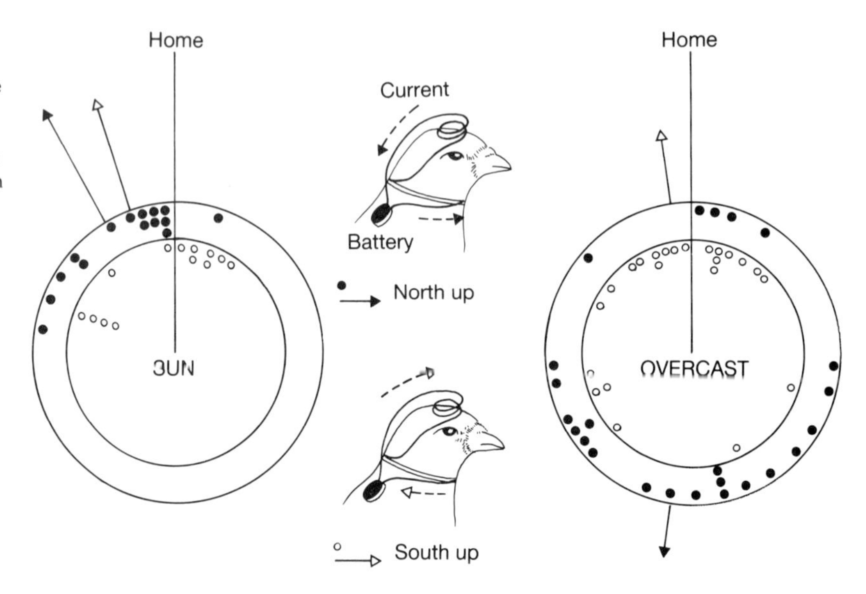

Figure 4.9 The orientation of homing pigeons is influenced by changes in the magnetic field. The dots represent the angles at which pigeons flew off from the release site: the direction home is straight up; straight down represents 180° away from home. There are two magnetic conditions, which do not affect orientation on sunny days (left) but do on overcast days (right). (After Baker)

Magnetic polarity can clearly provide one co-ordinate of a map sense; various possibilities have been offered as to how magnetism might provide a second co-ordinate, to complete the map. We shall not discuss the physical details here, only the evidence on the magnetic sense of pigeons. It is certainly true that pigeons with bar magnets attached to their heads navigate incorrectly on overcast, but not on sunny, days (Figure 4.9); which suggests both that pigeons have a magnetic sense, and that it may be used by pigeons as a back-up compass sense when the sun is invisible. Moreover, pigeons navigate less accurately when the earth's magnetic field is disturbed (for instance by solar flares); — but this evidence is inconclusive, because other factors may be correlated with these natural, not experimental, disturbances. Whether pigeons use a magnetic compass is still controversial, but a magnetic map sense is altogether more speculative; at present it is merely a theoretical suggestion. It does raise the question of how pigeons detect the magnetic field. No magnetic sense organ has been identified, but two hypotheses have been put forward. One is that the light-sensitive pigments of the eye could also act as magnetic sense organs. The other is that the molecule magnetite may act as the magnetic sense, and traces of magnetite have indeed been found in bones in the brains of pigeons and other species including humans. This research is still at an early stage. The magnetic hypothesis of homing is not yet confirmed.

The olfactory hypothesis has not been confirmed either. This hypothesis has yielded contradictory results. It would require pigeons (like salmon) to remember a home odour, or a 'landscape' of odours around its home, and to orient with respect to them. The hypothesis has been championed by F. Papi. In his most persuasive (but unrepeated) experiment, he trained a group of pigeons in a loft into which he experimentally made the wind blowing from the south to smell of olive oil, and the wind from the north to smell of turpentine (another group had the directions of the smells reversed). He then took the pigeons south, and placed a drop of olive oil on the bills of half of them, and a drop of turpentine on the bills of the other half. Those with turpentine, for instance, flew southwards if they had been trained in a loft with the smell of that compound blowing in from the north, and northwards if they had learned to smell it from the south. It seems they were using odours to guide them. In other experiments, Papi has shown that pigeons do not home accurately if their nostrils are blocked, or if their olfactory nerve is cut, or if they are experimentally distracted by a strong smell; but other experimenters have failed to repeat these results and it has proved impossible to train pigeons to make the relevant olfactory distinctions (which is a standard, and powerful, method of testing for sensory abilities — p. 43. In the face of this evidence, no conclusion should be drawn either way at present. The olfactory hypothesis, like its competitors (or complementary hypotheses), is not firmly substantiated. We cannot conclude with confidence whether pigeons have a map sense at all, and if so whether they use the sun or magnetic co-ordinates. Their ability to use landmarks and home cues is not in doubt, although whether they use olfactory cues as well as visual ones is still uncertain. Pigeons can home in familiar areas by remembered cues and a compass sense that is primarily derived from the position of the sun. Whether they can home from completely unfamiliar release points, and, if they can, how they do so, remains a puzzle.

4.3 Summary

Animals migrate in order to find environments superior to the one they are in. When environments change, migration may become advantageous. Migrating animals need a mechanism to locate superior environments. Wildebeest rely on sensing distant rainfall, which precedes environmental improvement. The predictable seasonal changes are anticipated by birds through the changes of day length. Emigration from overcrowded areas can result in the regulation of population densities.

Bee-killing digger wasps find their way home by memorizing the landmarks around their burrow entrance. Salmon find theirs by the perhaps memorized odour of their riverine birth-place. The most versatile home-finders are pigeons, which may be able to find their way home from unfamiliar starting points. They can navigate home even when taken to the release point under such conditions as would make learning the way out impossible. They probably navigate home by means of 'compass' and 'map', although the need for a map sense has been challenged. They use the sun, and perhaps the earth's magnetic field when the sun is invisible, as a compass. Their map sense is unknown, although the sun, the earth's magnetic fields, and olfactory maps have all been suggested and experimented on.

4.4 Further reading

Tinbergen (1958) describes his delightful experiments on *Philanthus*, only one of which did I have space to explain. Stabell (1984) critically reviews the subject of salmon homing. The mechanism of navigation is the subject of a large literature to which Gould (1982) and Baker (1984) and the contributors to Schmidt-Koenig and Keeton's book (1978) provide introductions. Baker (1982) has also written an introduction to the whole subject of migration in animals.

5 / Eating and not being eaten

5.1 Eating

The abstract principle of feeding is very simple: it is to find and catch food for yourself, while not being caught as food by another. The exact techniques used by different species are determined by the nature of their diet and predators, or, in other words, by their ecology. (Ecology is the science that deals with the relations of species, and communities of species, with their environment.) To understand the diversity of behavioural adaptations for feeding, therefore, we must first understand the ecological divisions of feeding types. For, although the methods by which different animal species obtain their food are about as diverse as the number of animal species, there are certain broad patterns to be seen. A first distinction is between herbivores and carnivores. The energy input to the world comes from the sun — plants grow on the sun's energy by photosynthesis — herbivores feed directly on the plants, carnivores feed on the herbivores. The chain has further steps. Carnivores feed on other carnivores (which make up 10 percent of the diet of the leopard, for example), scavengers feed on the dead carcases of all kinds of animal, and most important, the decomposing fungi and bacteria return the nutrients from dead bodies to the soil, from whence they can be re-used by plants.

We have some understanding of the relative proportions of the different ecological types. It was first observed early in the century that there is usually only about 10% as much energy (in the form of animal matter) at one level of the food chain as at the next level down; at any one place, about 10% as much energy will be contained in the carnivores as in the herbivores. The reason was first realized by such early ecologists as the Englishman, Charles Elton and the American Raymond Lindemann; it is the energetic inefficiency of transferring food from one level to the next. For energy to be converted from zebra into lion, the lion has to chase and kill the zebra, and then digest the zebra meat. All of these processes burn up fuel and energy is therefore lost in the transfer.

The behavioural adaptations of herbivore and carnivore will obviously differ. The herbivore has no difficulty in catching its food, although it may be difficult to select the most edible parts of the plants, which are frequently tough, indigestible, short of nutritive value, and stuffed with poisons. The carnivore will at least have to be mobile to catch its moving prey. The distinctions at this high level are relatively unilluminating: we can better see how feeding ecology and feeding behaviour interact if we examine more closely a smaller group of species. Let us consider, for example, the feeding habits of the five main species of large herbivorous mammals that inhabit the Serengeti Plains of East Africa.

5.2 Feeding in group-living herbivores

Buffalo, zebra, wildebeest, topi, and Thomson's gazelle live together in huge groups which together make up some 90% of the total weight of mammals living on the Serengeti. At first sight the five all appear to live on the same species of grass, herbs, and small bushes. The appearance, however, is illusory. When Bell and his colleagues analysed the contents of the stomachs of four of the five (they did not study buffalo), they found that each species was living on a different part of the vegetation. These different parts of the vegetation differ in their food qualities: lower down, there are succulent, nutritious leaves; higher up are the harder stems. There are also sparsely distributed, highly nutritious fruits, and Bell and his colleagues found that only the Thomson's gazelle eats much of these. The other three species differ in the proportions of lower leaves and higher stems that they eat: the zebra eats the most stem material; the wildebeest the most of the leaves; the topi is intermediate.

How are we to understand their different feeding preferences? The answer seems to lie in two associated differences among the species, one in their digestive systems and the other in their body sizes. The digestive systems can be divided into the non-ruminants (the zebra, which is like the horse) and the ruminants (wildebeest, topi, and gazelle, which are like the cow). Non-ruminants cannot extract much energy from the hard parts of the plant; however, this is more than made up for by the fact that food passes much more quickly through their guts. Thus, when there is only a short supply of poor quality food, the wildebeest, topi and gazelle enjoy an advantage. They are ruminants and have special structures in their stomachs (the rumen), containing special micro-organisms which can break down the hard parts of the plants. Food passes only slowly through the ruminant's guts because ruminating, digesting the hard parts, takes time. The ruminant continually regurgitates food from its stomach to its mouth to chew it up further (that is what a cow is doing when 'chewing cud'). Only when it has been chewed up almost to a liquid can the food pass through the rumen, and on through the gut. Larger particles cannot pass through until they have been chewed down to size. Therefore when food is in short supply, a ruminant can last longer than a non-ruminant because it can extract more energy out of the same food. The differences can partially explain the eating habits of the Serengeti herbivores. The zebra chooses areas where there is more low quality food. It migrates first to unexploited areas and chomps the abundant low quality stems before moving on. It is a fast-in/fast-out feeder, relying on a high throughput of incompletely digested food. By the time the wildebeest (and other ruminants) arrive, the grazing and trampling of the zebras will have worn the vegetation down. As the ruminants then set to work they eat down to the lower, leafier parts of the vegetation. All of which fits in with the differences of stomach contents with which we began.

The other part of the explanation is body size. Larger animals require more food than smaller animals, but smaller animals have a higher metabolic rate. Smaller animals can therefore live where there is less food, provided that it is of high energy content. That is why the smallest of the herbivores, Thomson's gazelle, lives on fruit, which is very nutritious but too thin on the ground to support a larger animal. By contrast, the large zebra lives on the masses of low quality stem material.

The differences in feeding preferences lead, in turn, to differences in migratory habits. We have seen (p. 84) that wildebeest follow, in their migration, the capricious pattern of local rainfall. The other species do likewise. But when a new area is fuelled, by rain, for exploitation, the mammals migrate towards it in an orderly pattern. The larger, less fastidious feeders, the zebras, move in first; the choosier, smaller wildebeest come later; and the smallest species of all, Thomson's gazelle, arrive last (Figure 5.1). The later species depend on the preparations of the earlier, for the action of the zebra fits the vegetation for the stomachs of the wildebeest and gazelle.

If we are to understand the feeding habits of the species, therefore, we must consider it in relation to the whole ecology of the species, and its relations to other species. Behaviour is an inseparable part of a whole system, made up (in this example) of body size, gut morphology, and the habits of associated species.

Figure 5.1 The main species of large herbivore of the Serengeti Plains, East Africa, migrate into an area after rainfall in a predictable order: after the rain, the grass grows; zebras then arrive first, followed by wildebeest, and Thomson's gazelle. The order of migration can be explained by the different diets of the different species. (After Bell)

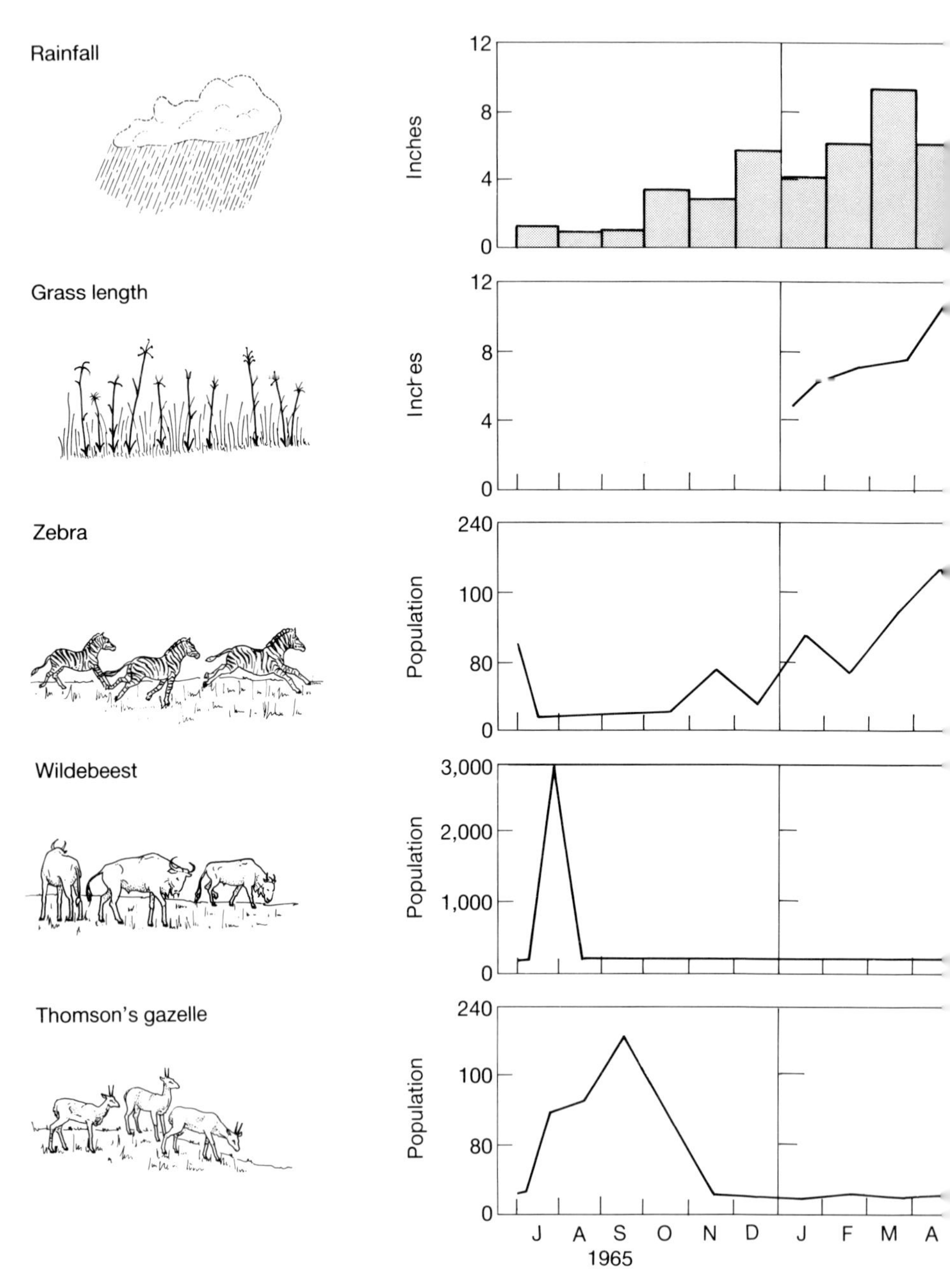

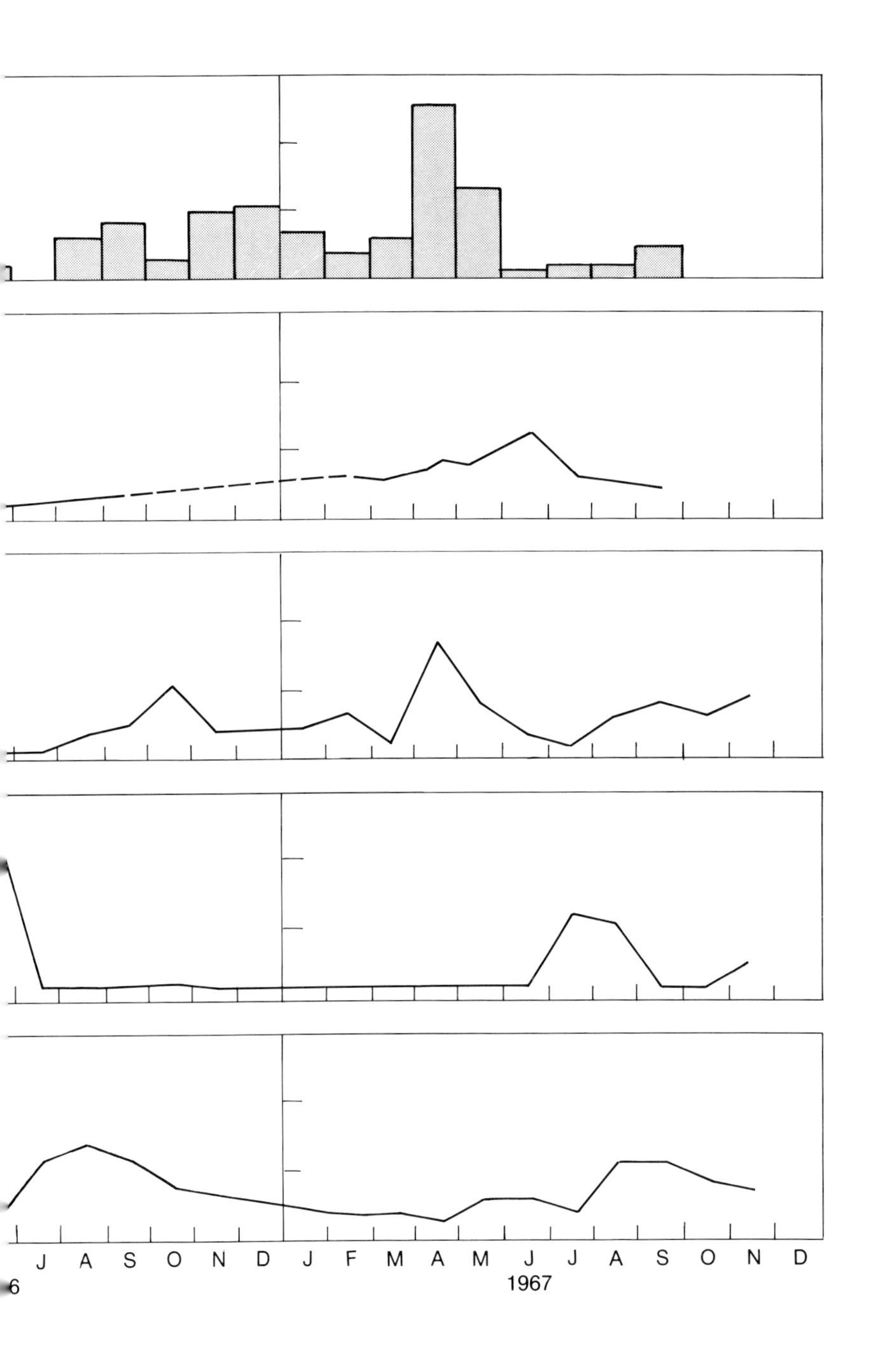
J A S O N D J F M A M J J A S O N D
6
1967

5.3 Recognizing food

Given the diet of a species, determined by its ecology, the individual members have two behavioural problems, the best solutions to which will depend on their diet. Food must first be recognized, and then caught. We shall consider these two in the next two sections. First how is food to be found? The environment of an animal contains all sorts of things, some of which are edible and some of which are not. The problem for the hungry animal is to distinguish the former from the latter, and only to put the edible things in its mouth. What is food for one species is often poisonous, or indigestible, for another. Each species must recognize its own kind of food. The complexity of the problem depends among other things on whether the species is a specialist feeder which eats only one or two types of prey, or whether it is an omnivore which eats many kinds of prey.

Let us consider first a fairly specialist feeder, the toad. Toads mainly eat small, dark, flying insects. They stick out their tongue and snap at any insect which flies by. J.-P. Ewert and his colleagues have studied in detail how the toad recognizes its meal. The toad does not recognize insects as such, but recognizes small, dark, moving objects. If a piece of dark paper about ½ inch in length is moved where a toad can see it, it will snap at it just as if it were an insect (Figure 5.2). This recognition system works perfectly well for a toad because any small, dark, moving object in its natural environment *is* an insect. Only in the laboratory is it fooled by bits of paper. Ewert and his colleagues have also worked out where, in its nervous system, the toad recognizes small dark moving objects. They found neurons in the retina of the eye which respond differently according to the size of dark objects moving in the field of view. There are three classes of sensory neurons which respond to three different classes of objects. The visual system of a toad is a world of moving dark objects of various sizes and shapes; its gastronomic choice is to put all the objects of a certain size in its mouth.

The toad's response to different sized objects is not much affected by learning; to find cases where learning has an influence we must turn to animals with more catholic diets. The first learning phenomenon to consider is a perceptual one, called a search image. Forming a search image means learning to see something which had not previously been seen. It often occurs in humans with photographs of camouflaged animals. To begin with you cannot see the animal at all, but after you have first noticed it, it becomes

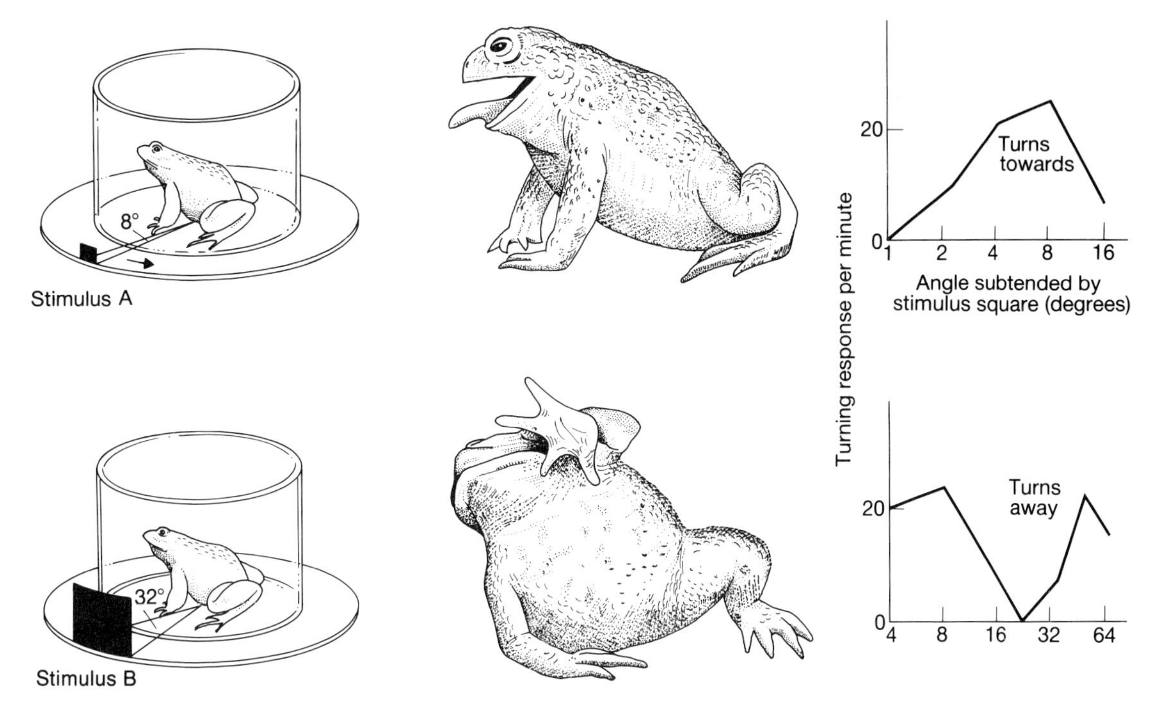

Figure 5.2 Toads recognize small moving objects as food. In the experiment illustrated at the top, the toad's response to a small black object revolving in its visual field is measured. The graph on the right reveals that the response depends on the escape. From Ewert 'The neural basis of visually guided behavior'. Copyright © 1974 by Scientific American, Inc. All rights reserved.

obvious. You have learnt to see it. It is more difficult to know whether an animal goes through the experience of at one moment not being able to see a food item, but then being able to see it at the next. A human can tell you of this experience verbally; with animals you can only study what they do. If an animal does not eat a food item at one moment but does later, it may only be because it was not hungry before, not because it has learned to see it. An experiment by M. Dawkins clearly demonstrates the formation of a search image. She worked with domestic chicks, feeding on rice grains. She dyed some of the rice grains a different colour from the background, these grains were easy to see: she dyed other grains the same colour as the background, which made them difficult to see. The chicks pecked the conspicuous rice grains, which proves that they were eager to eat rice. However, there was a delay of a few minutes from the start of the experiment before the chicks ate their first camouflaged rice grains. After eating one camouflaged grain, they rapidly ate more of them. To be precise, there was an average delay of 66 seconds from eating a visible grain to eating a camouflaged one, but a delay of only 6.7 seconds from eating one camouflaged grain to eating another camouflaged one. The chicks learned to see camouflaged grains; they formed a search image for them.

Learning becomes even more important in the feeding of so omnivorous an animal as the rat. Rats will initially sample, in small quantities, almost anything; but if they find what they have taken makes them sick, they avoid eating anything like it again. This 'one-trial' learning, however, does not operate in the perceptual system of the rat, but in its decision-making machinery. It is easy to see that the simple recognition system of the toad would be utterly inadequate for an animal that must be able to distinguish between a large number of potentially edible objects.

5.4 Hunting food

5.4.1 Foraging on immobile prey

Animals that feed on static food have to move around to find it; and, as we shall see, the best course of movement depends on the distribution of the food items in space. If the prey can themselves move, the predator will have to be able to move faster, or with sufficient stealth that it can catch its prey without a long chase. It may pay to hunt in groups; prey can then be surrounded rather than chased, and large species of prey can be subdued by a number of smaller predators. Other animals, such as spiders, caddis flies, mantises and angler fish, do not move to catch their mobile prey. They 'sit and wait', and ambush prey that move to them.

There is a general principle of feeding that the animal's pattern of movement — hunting, foraging, or searching — should fit the distribution of prey. This is now a flourishing area of research, but we shall pick on only one illustration. J. N. M. Smith put out pastry 'caterpillars' for thrushes, and watched how the birds adjusted their movements to the arrangement of prey. The prey could be placed with regular spacing, or at random, or clumped in groups. Smith found that the thrushes moved differently when the prey was clumped from when it was spaced out (Figure 5.3). Thrushes move in discrete jumps, of which Smith could measure two properties: their length, and the angle of turn between two jumps. When food was clumped, a thrush, after finding a food item, turned through a sharper angle than when it was spaced out, but kept the length of each jump constant independent of the food distribution. The turning angles used by the thrush make sense. When food

Figure 5.3 Area restricted search. Birds can adjust their movements according to the distribution of food. At the left the movements of a ground-feeding bird, such as a song thrush *Turdus philomelos*, are shown for a bird that has learned that its food is clumped. It strides more or less straight ahead until it finds an item of food, and then changes its movements, now turning through tight angles. The movements illustrated on the right are those of a bird that has learned that its food is spaced out. It keeps walking straight ahead even after it has found a bit of food.

is clumped in groups, the best method of search is to move straight ahead when not in a group, but to turn through tight angles once one item has been found, in order to find the rest of the group. However, when food is spaced out, the bird is unlikely to find another food item nearby after finding one, and it is best to keep moving on. The change in behaviour for clumped food items is called area-restricted searching. It is an adaptation to increase the rate of finding food.

5.4.2 Group-hunting carnivores

Carnivores that live on mobile prey by no means all live in groups. But we do not have space to consider all feeding methods, and will pick on group hunters as one example of the active carnivorous habit (Figure 5.4). It will incidentally allow us to consider the question of what advantage group-living has. A pride of lions hunting down a prey animal, such as a zebra, is one of nature's more awesome spectacles. A complete pride might contain two males, seven females, and some cubs, but the hunt is a females' business. The males stay behind, for their showy manes would only disturb the hunt whose technique is stealth.

The hunt starts with the lionesses hunting as a group for their prey. When they spot some promising zebras, or antelopes, they spread out into a line.

Figure 5.4 A pride of lionesses on the Serengeti Plains, Kenya. On large prey, lionesses hunt more efficiently in co-ordinated groups. (Photo: Heather Angel)

They now move slowly, stealthily towards their unknowing victims. Soon one lioness is close enough to a zebra to attack. In the ensuing panic zebras run in all directions, many of them right into the paws and jaws of another waiting lioness. The lionesses also co-operate in killing prey. If the prey is large and dangerous, such as a buffalo, the attack of several lionesses is the safest method of execution.

Not all group hunts are successful. In fact, of the group hunts watched by Schaller on the Serengeti Plains, only about 30% were successful. Lionesses hunting in groups are, however, more successful than lionesses hunting alone. Lionesses often hunt alone, but only about 15% of these hunts are successful. The advantage to carnivores such as lions of hunting in groups is that they can catch more food than if they hunted alone. They can also catch kinds of prey that they could not catch by themselves, such as buffalo. Carnivores can also adjust the size of the hunting party to the kind of prey being sought. Hyenas, to take another example, hunt in groups for zebra, but hunt singly for the smaller Thomson's gazelle.

Lions are strong but cannot run for a long time. A zebra, or antelope, could easily outrun a lion, and lions must therefore rely on stealth and the surprise attack; carnivorous dogs, by contrast, hunt by running down their prey. The African hunting dog and the wolf employ similar hunting techniques. Wolves hunting moose, or hunting dogs wildebeest, start by charging several of the prey in a group. The prey easily runs away, but the dogs have a chance to select a vulnerable individual. They single out an old, young, or infirm animal and only then start the chase. During the chase, the dogs of the pack may take turns in leading. When the kill comes they usually all join in to wear their victim down. They are flexible, however, and employ other techniques when appropriate. They pay particular attention to prey that straggle or leave the herd. One dog may then try to get between the herd and the straggler, and try to drive it towards the rest of the pack.

5.4.3 A note on domestication

Domestic dogs are descendants of wolves, to which they show many similarities of appearance and behaviour. Their social habits have fitted them to human society. The exact reasons why dogs were first domesticated by humans are now lost in the past, whether it was the agreeable companionship of an animal that apparently respects its master much as a dominant member of its hunting pack, the exceptional sensitivity of their whole body surface, their uses in agriculture, or, most likely, a combination of all such factors. Undoubtedly their ancestry rendered dogs particularly good pets. We still

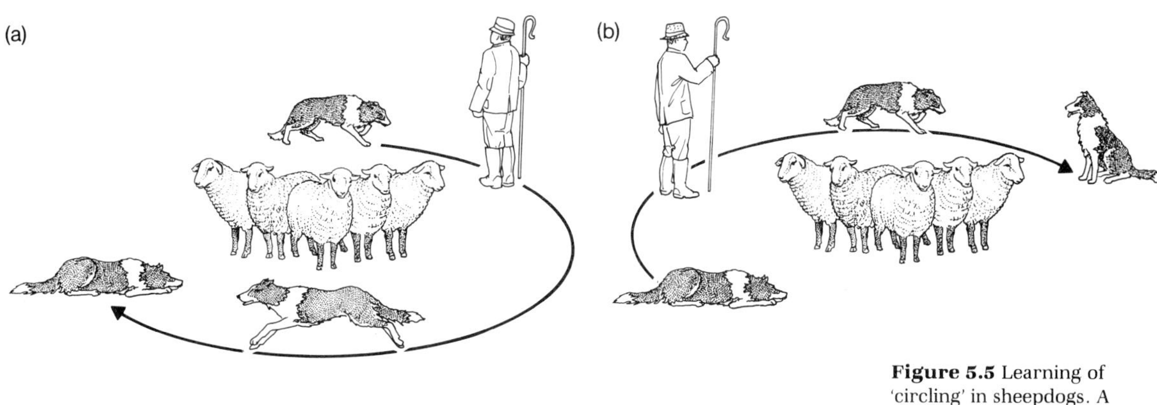

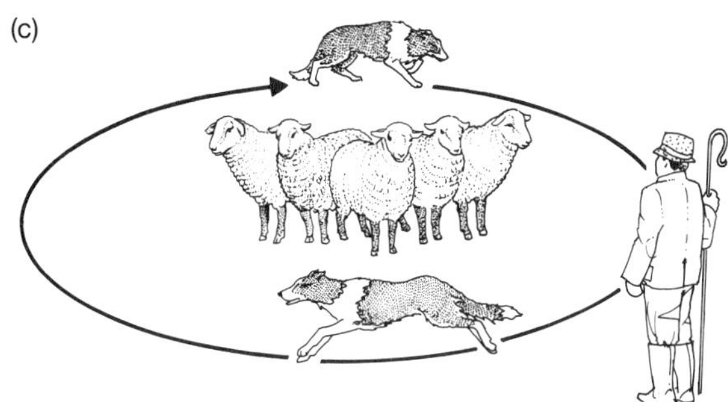

Figure 5.5 Learning of 'circling' in sheepdogs. A naive dog will spontaneously run round to the other side of the flock. It can therefore be taught to circle if (a) the shepherd first makes the dog run round to the other side of the flock, and then (b) moves himself and repeats the exercise. Soon (c) the dog will encircle the flock by himself. (After Vines)

exploit the ancestral habits of dogs in their training. The hunting techniques we have just discussed for wolves and African dogs are made use of in the training of sheepdogs. For instance, young and untrained sheepdogs will often spontaneously run round to the other side of a flock of sheep and try to drive them towards the shepherd. It is only a small step from this to teach the sheepdog to 'circle', to run round the flock and keep the sheep tightly bunched together (Figure 5.5). Sheepdogs are also particularly alert to stray sheep and can easily be taught to drive stray sheep back to the shepherd. If we imagine that the sheepdog is treating the shepherd as a fellow hunter, then circling and retrieving stray sheep both manifestly resemble the ancestral canine hunting behaviour. Moreover, the most difficult tricks to teach a sheepdog are those most removed from its ancestral hunting skills. Sheepdogs are difficult to teach to drive the flock away from the shepherd — the dog has to be restrained from its desire to circle round and drive them back. Sheepdogs are also difficult to teach to leave stray sheep they have gathered, to go and gather more.

The learning abilities of the sheepdog are closely related to its ancestry. It learns most easily what comes naturally to it. This is a general principle in teaching tricks to animals. Circus animals, for another example, learn tricks most easily if those tricks are a simple extension of the animals' natural behaviour patterns.

5.5 Avoiding being eaten: active evasion

Any property of an organism that reduces its chances of being taken by a predator will be favoured by natural selection, as it will increase the organism's chance of survival. The resulting anti-predator adaptations are very diverse. In the following four sections we shall examine five different habits used by animals to avoid being eaten: potential prey may actively flee their predators, or they may stay still and try to be invisible, or they may stuff themselves with sickening chemicals and advertise their unpalatibility with bright 'warning colours', or they may mimic the warning colours of others, and finally, in some circumstances, an animal may make itself less likely to be eaten by living in a group.

Active flight is used by many animals to escape predators, and a particularly elegant study has been made on noctuid moths by Roeder. Noctuid moths are eaten by bats, and have evolved a special pair of ears to warn them of approaching danger. There is one ear on each side of the thorax, and each has a simple structure; two nerves connect each ear with the thoracic ganglion (which is the nearest mini-brain). They are sensitive to the high pitched squeaks used by bats to find their prey, and they have the advantage of the bat, in that the bat emits very loud blasts in order to detect a faint echo. The moth can hear the bat from a greater distance than the bat can pick up the echo from a flying moth; to be precise, a moth can hear a bat about 100 feet away, whereas a bat can detect a moth at a range of less than 8 feet. The moth, moreover, can tell whether the bat is to the right or left (because it has an ear on each side) and whether it is approaching or moving away. A bat approaching a moth will sound louder and louder as it comes close, and the moth is sensitive to loudness. Bats do not fly in the same direction for long. Therefore, if a moth hears a bat approaching about 100 feet away, its best policy is to fly off in the other direction. That way it may get out of the bat's flight path before it enters the detection range. Once a bat has detected a moth it has the advantage, because bats can fly much faster than moths. The moth's surest means of staying alive, therefore, is not to be detected; and it has the advantage of advance warning to keep out of the way.

A bat may appear suddenly out of the dark close to a moth. It is then useless for the moth to flee, because it will have been detected and the bat can fly faster than it. When a moth detects a loud bat sound, indicating a bat less than about eight feet away, it puts a different escape tactic into action. It flies in wild loops and spirals, and dives to the ground, a course of flight designed to make it as difficult as possible for the bat to catch it. (The erratic flight may be produced by the moth by just switching off its steering mechanism. Then even the moth will not know where it is going: and the most effective means of confusing someone else about where you are going is not to know yourself.) The moth has two escape responses. If it hears a bat afar, it turns and flees; if it is surprised by one nearby, it goes into a crazy flight. It uses its hearing sense to decide which response is appropriate.

5.6 Camouflage

The nochid moth's defence is to seek escape in active flight. The opposite defence is to sit dead still and try to be invisible. Such is the method of camouflage in which a species evolves to resemble its background. Camouflage is of course an adaptatin of appearance and colouration, but the most exquisite artistry will be wasted if the animal's behaviour is not suited to the camouflage. The world is a patch-work of different colours: the animal is only camouflaged if it settles in the right place. Consider the European grasshopper *Acrida turrita*. It comes in a green form and a yellow form. In nature the green form lives in green places and the yellow form in yellow and brown places, with rare exceptions. In a simple experiment, the German ethologist S. Ergene gave yellow and green grasshoppers a choice between yellow and green backgrounds. The green grasshoppers fittingly tended to go and settle on the green backgrounds, and the yellow grasshoppers on the yellow.

The North American moth *Melanolophia canadaria* faces a more difficult problem in lining up with its background. It has striped wings and lives on the bark of trees. It must line its stripes up with the lines of the bark if it is to be camouflaged. In an experiment, T. D. Sargent allowed the moths to sit on cylinders that had regions of vertical stripes and regions of horizontal stripes. If the stripes (which were made of black tape, stuck on a white surface) could be felt by the moths, then the moths usually lined up correctly. When Sargent covered up the stripes and surface with a transparent film, the moths no longer lined up correctly. The moths must be relying on the feel of the surface that they have to line up on. In nature they will be able to feel the stripes of their background, and ensure that they settle in a camouflaged posture.

5.7 Warning coloration and mimicry

Some animals protect themselves against being eaten by containing poisonous or sickening substances. Some such animals make their own poisons; others take them from sources in their environment. The wings of the monarch butterfly, for example, contain powerful heart-stopping poisons called cardiac glycosides. The monarch eats the poisons as a caterpillar, when its food plant is the asclepiad, or 'milkweed', which contains cardiac glycosides. The caterpillar is not harmed by the poisons; it just stores them, and they are then retained by the adult.

For the behavioural problem of defence by poison, we must turn from the prey to the predator. If the defence is to work, the predator must learn not to eat poisonous animals, because natural selection will not favour a trait, by which an animal, after it is dead, makes its attacker sick. If the trait is to evolve, it must ensure survival. The tactic used is to enable predators to learn to recognize sick-making prey, in which case they will avoid them. It is a skill predators will readily learn. When a bird eats a monarch, as an adult or caterpillar, it will be violently sick within minutes, an experience it would learn not to repeat. The bird's problem then is to distinguish sickening from edible prey. Now, the colours of the monarch are bright gold, and ethologists suspect that the bright coloration evolved to make the monarch more memorable to birds. It is therefore called 'warning coloration'. Experiments have shown that predators learn to avoid sickening prey, J. V. Z. Brower, for instance, offered the monarch butterfly as food to the bird called the Florida scrub jay. On their first meal of monarch the jays were violently sick; but after only a few trials they had learned not to eat monarchs, though they continued to eat other, tasty food. What has not yet been conclusively demonstrated is that birds learn to avoid brightly coloured sickening prey more quickly than equivalent but duller coloured prey. It is necessary for the theory that they should, for otherwise poisonous prey might just as well be dull as brightly coloured. Few ethologists would doubt, however, that such an experiment would be successful.

The early stages in the evolution of warning coloration pose a paradox. The population of monarchs clearly benefits from teaching each generation of birds not to eat them. They benefit by being eaten less. The problem is that some butterflies have to be eaten to teach the birds to begin with. In evolutionary terms, it is no consolation to the dead butterflies that some others are benefitting from their death. They are dead; they have failed to reproduce; natural selection has worked against them. When warning coloration first appeared in evolution, it would have been a rare, minority characteristic. Those rare, brightly-coloured butterflies would have been conspicuous to predators, and therefore eaten. One would expect natural selection to have eliminated the characteristic. How could natural selection favour an increase in its frequency? The question has not been satisfactorily answered, though there are some possible answers. One possible answer is that the family relatives of the eaten butterfly may benefit from its death. Members of the same family tend to resemble each other; if one were warningly coloured, many others would be as well. If one member were killed, the rest of the family would benefit from the lesson taught to the predator.

However that may be, the learning by predators to avoid prey of a particular appearance makes possible the evolution of another strategy: mimicry. Predators, we have seen, learn not to eat monarchs because they are sickening. But another butterfly, the viceroy, inhabits some parts of the monarch's geographic range, and is not poisonous to birds. It also looks very like a monarch. It can presumably survive, even though it is edible to birds, because birds take it for a monarch, and avoid it. An experiment suggests that this is so. Brower divided eight Florida scrub jays into two groups of four each. She first offered monarchs to four 'experimental' birds and viceroys to four 'control' birds; she recorded how many times the birds avoided the butterflies or pecked at them. The experimental jays should learn that the colour pattern is associated with unpalatable food; the control jays should not. Brower then offered viceroys to both classes of jays and recorded whether they avoided or pecked at them. The experimental birds now tended to avoid the viceroys, and the control birds to peck at them (Table 5.1). The result suggests that Florida scrub jays may indeed avoid mimics after they have learned to avoid the sickening model.

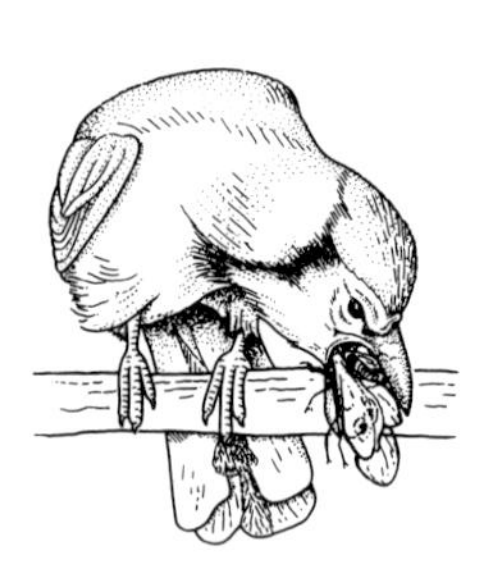

Table 5.1 Experimental demonstration of mimicry. Brower had offered four Florida scrub jays ('control' birds) palatable viceroy butterflies and four other Florida scrub jays ('experimental' birds) sickening monarch butterflies. The monarch and viceroy look alike. Brower then offered viceroys to both classes of jays and recorded the number of trials in which they avoided the butterflies or pecked at them; the numbers in the Table give her results. The difference is statistically significant. (Slightly simplified from Brower, 1958.)

	Control birds				**Experimental birds**			
Number of trials in which jay:	C–1	C–2	C–3	C–4	E–1	E–2	E–3	E–4
Avoided	0	0	9	1	14	12	12	4
Pecked	25	25	16	24	2	1	3	12

Mimicry in which a relatively tasty prey mimics another poisonous species is called Batesian mimicry after its discoverer, the nineteenth century British explorer H. W. Bates. Another nineteenth century explorer (like Bates, in South America) was the German, Fritz Müller. Müller discovered another kind of mimicry, now called Müllerian mimicry, in which all the mimicking species are poisonous. All the species in a Müllerian mimic group benefit when an individual of any one species is eaten. The most extraordinary development of Müllerian mimicry is that of the South American butterflies in the genus *Heliconius*. Two species, *H. erato* and *H. melpomene*, are Müllerian mimics. In any one place the two species resemble each other, but in different places they have formed 'geographic races' (Figure 5.6). The butterflies look different in different places; but the two species always change in exactly the same way. Thus two individuals of different species

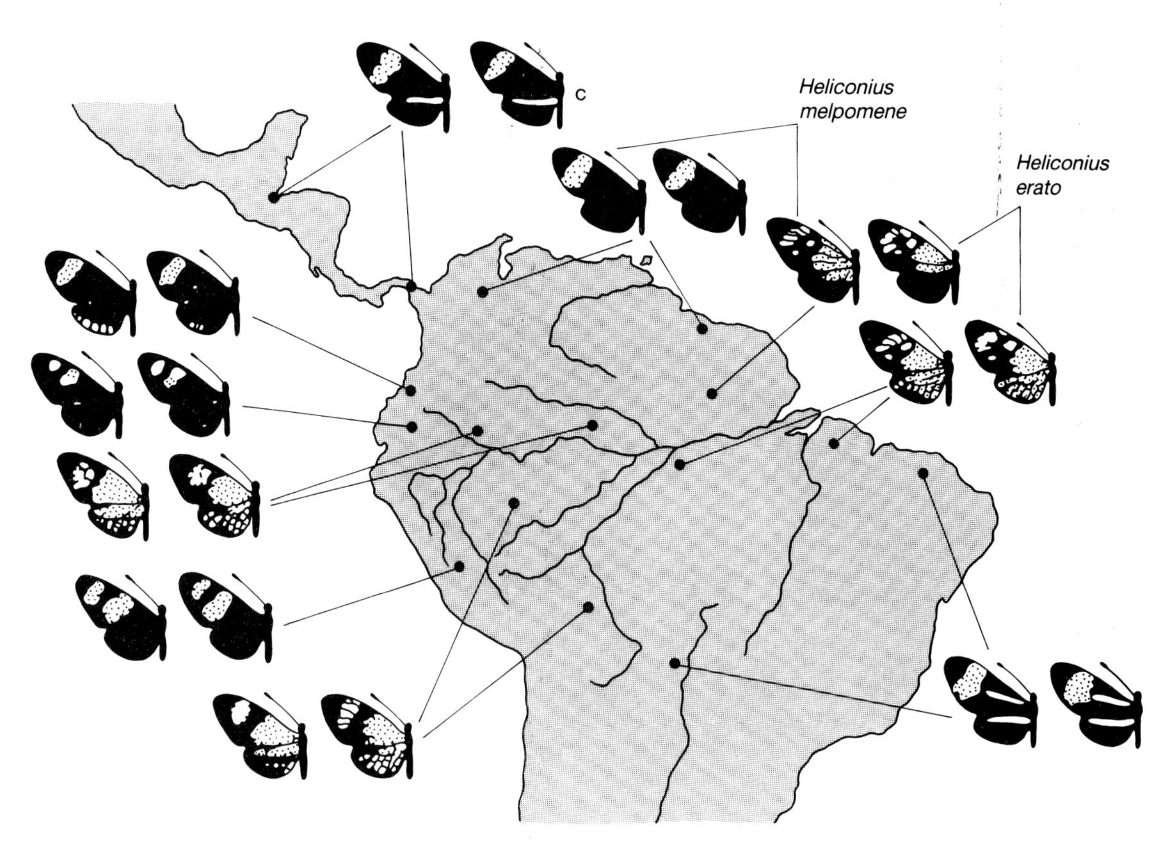

Figure 5.6 The two butterfly species *Heliconius melpomene* and *Heliconius erato* form remarkable parallel mimicry 'rings' in central and South America. In any one place the two species mimic each other, looking much the same; but in different places the members of a species differ: both species vary geographically in the same manner. *Heliconius* butterflies are poisonous to birds. (After Turner)

from the same place look more alike than two individuals of the same species from different places. In any one place, the two species gain by resembling each other, because predators will treat them both as the same kind of prey; but few predators move far enough for there to be any advantage to the *Heliconius* in looking the same in distant places. All that matters is to educate the local predators, and that can be done with different colours in different places. The particular colour pattern of a *Heliconius* presumably does not matter much, so long as it is memorable to the local birds. Their colour pattern will then protect them from predation.

5.8 Fish schooling

In many species of fish, individuals swim together in large groups called schools. The habit of moving around in large groups is much commoner in fish than in other kinds of animals. Within fish, it is commoner among species that are small in size, and active swimmers; and even within a species, small individuals may school whereas large ones do not. Small tuna, for example, swim in schools, but when they grow large they become solitary. Very few of the large predatory species of fish swim in schools, though barracuda are an exception. The observation that large, predatory species of fish do not usually swim in schools might lead us to suggest that the advantage of schooling is in defence against predators, for large predatory fish are not themselves subject to predation.

Schooling might offer any of several kinds of protection. One is that a predator is less likely to detect fish prey if they form a school than if they spread out. There is an analogy here with ships in wartime, which sail in convoys to reduce the numbers lost to submarines. If all the ships sail separately it is more likely that at least one ship will stray into the field of view of a submarine than if the ships sail in a tight convoy; a convoy may slip by without being detected. Even once a predator has found a school, the prey fish are better off in the school than they would be alone, as has been proved by the following experiment. A prey species, such as the bleak or dace, is put in tanks either singly or in groups of up to 20; the groups will form schools. The experimenter then puts a predatory species of fish, such as the pike, into each tank. The result is that the predator catches more prey per unit time when feeding on single fish than when feeding on schools. The reason is that in a school the predator is confused by the multiplicity of moving prey items: it cannot concentrate on one without being distracted by another. The pike is perhaps in a similar difficulty as a human who is trying to hit a moving tennis ball when someone throws a second ball across his visual field; it makes it more difficult to hit the ball. The predator, moreover, has to concentrate when there are hundreds of whizzing prey. Another experiment supports the same theory. If predators find it difficult to concentrate on prey in schools, then they should become more efficient if some members of the school are made easier to concentrate upon. If an experimenter makes a few fish in the school distinctive, the predator should take more notice of them. And, indeed, when Indian ink was daubed on a few minnows, which were then put among a school, the

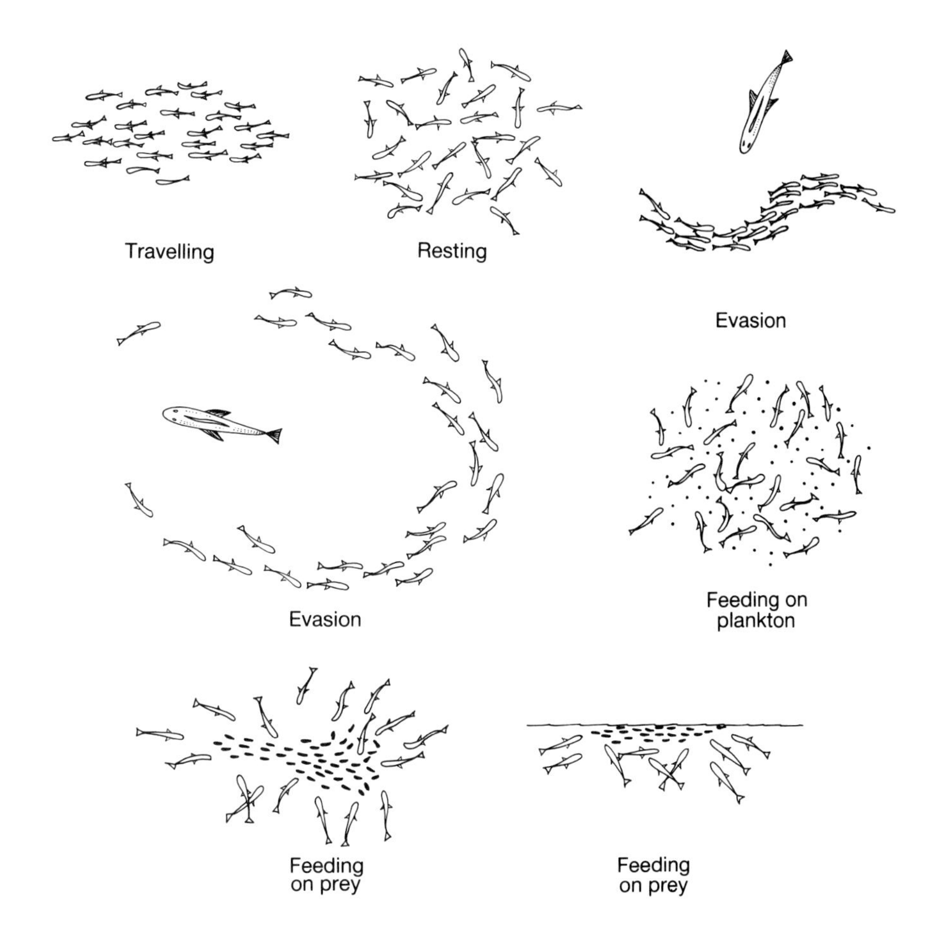

Figure 5.7 The arrangement of fish in a school depends on their environment. They align themselves differently according to whether they are travelling, resting, feeding, or avoiding being fed on. The school becomes less co-ordinated when the fish are resting, or feeding. (After Wilson)

painted minnows were disproportionately taken by predators. It was easier for the predator to keep its eye on them: in the normal school the predator is visually confused.

Schooling fish also take active steps to reduce the chance of predation (Figure 5.7). Schooling sticklebacks and catfish, for example, close ranks on the approach of a predator. Other species deploy more spectacular 'set piece' avoidance tactics. When a school is attacked, the fish may 'explode' away from the point of attack, in a dramatically simultaneous rapid movement. It may take less than a tenth of a second, yet the fish do not bump into each other. Another defensive manoeuvre is for the school to break in two in front of the predator, and then reform after the two sides have swum round behind the predator.

Schooling, then, is a complex adaptation to avoid being eaten. It has other advantages as well. Fish swimming in schools move faster and use less energy than fish of the same species made to swim alone. The reason for this is uncertain. The Atlantic spadefish provides evidence for one theory. The spadefish swim faster in schools because of a slime given off by the swimming fish. The slime reduces the drag experienced by the fish swimming behind, which can consequently swim faster. Schooling fish, therefore, swim faster as well as more safely than fish that swim alone.

We saw earlier, in the case of group-hunting carnivores, that living in groups can be an adaptation to increase the efficiency of hunting. It now appears that in other species the same habit can have an almost opposite function. Schooling in fish (and this is not the only example) serves to decrease the chance of being eaten. Group living therefore has different functions in different species.

5.9 Summary

An animal's behavioural adaptations must be understood in the context of its feeding ecology: of the diet it is seeking to fulfil. Each of the five main large herbivorous mammals of the Serengeti has a specific diet determined by its energetic needs and alimentary morphology, and possesses feeding preferences and migratory responses to obtain it. An animal must recognize, and catch, its food. The relatively inflexible toad recognizes food merely by the size of the dark moving objects in its visual field; its recognition mechanism is neurophysiologically understood. Some species learn to recognize their prey after practice, which is called forming a search image. Animal techniques of searching and catching food are fitted to their prey. Thrushes adjust their movements according to the spatial dispersion of their prey. Lions, dogs, and wolves hunt down prey by co-operative group mechanisms. These ancestral mechanisms, in the case of dogs, have influenced the process of domestication.

Animals may avoid being eaten by active flight, as do moths escaping from bats; by camouflage, which requires behavioural adaptations to fit in with the visual markings; by warning coloration, to teach predators to avoid sickening prey; by mimicry of successfully warning-coloured species; or by aggregation in groups, such as the schools of fish.

5.10 Further reading

Colinvaux (1978), discusses feeding ecologically. Bell (1971) explains his work on the herbivores of the Serengeti, and Ewert (1974) describes his work on food recognition in toads. M. Dawkins (1971) gives the details of her experiment on search images; rat learning can be followed up through Garcia *et al.* (1972). For the experiment on thrushes the reference is Smith (1974a,b), and the subject of search paths is discussed more generally by Pyke, *et al.* (1977) and by Krebs (1978); it is part of the general subject of 'optimal foraging theory', which is reviewed by Krebs, Stephens, and Sutherland (1983). The behaviour of group-hunting carnivores is described by Wilson (1975); see Schaller (1972) for lions. Vines (1981) was my source for sheepdogs. Roeder (1965) describes his work on moths; Turner (1977) the subject of mimicry; and Edmunds (1974) discusses defensive adaptations as a whole.

6 / Signals

6.1 Principles of communication

The subject of our remaining four chapters is social behaviour. All social behaviour is mediated and organized by communication, and before we come on to such social, and anti-social, topics as fighting, sex, and altruism, we should consider the means by which these interactions are controlled. We should discuss the principles of animal signals.

Ethologists never have been able to make up their minds about what they mean by communication and signalling. A definition, along the right lines but too broad, is that an animal has signalled when it changes the behaviour of another animal. The definition can be made more accurate, but not wholly satisfactory by specifying that the other animal must have changed its behaviour because it perceived the signal through its sense organs, and was not physically forced. On the former definition, pushing someone in a river would have to be called a signal because it certainly would change his behaviour: but this would be excluded by the definition requiring the influence of the signal to be mediated by the recipient's sense organs.

The criterion of behavioural change is necessary in order to recognize signals by external observation. Most signals have been discovered by simply watching the behaviour of interacting animals. If members of a species

Figure 6.1 Experiments are needed to confirm that any particular structure or behaviour pattern functions as a signal. Herring gull (a,b) and black-headed gull (c) chicks naturally peck at the tip of their parents' bills to beg for food. At the tip of the gull's bill is a red spot that was believed to be the signal to the chick to peck. Tinbergen presented various (d–f) more or less realistic model gull bills, with different coloured spots (compare e and f) to gull chicks. He measured the rate at which they pecked the different models, and confirmed that the red spot on the tip of the bill acts as a signal to the chick to peck, in order to beg for food. (Photos: Niko Tinbergen)

consistently perform an activity, such as running away, after other members have performed another recognizable activity, such as baring their teeth, then teeth-baring is probably a threat signal. The evidence of simple observation is not, however, perfectly convincing. It can only establish a correlation between two behaviour patterns; but a correlation can always be explained by both activities being caused by a third, unobserved activity. Experimental evidence is therefore more convincing. Tinbergen, for example, presented model bills of adult herring gulls (*Larus argentatus*) to the chicks of that species; the chicks responded by pecking at the red spot at the top of the bill, which strongly suggests that the red spot is a signal, meaning 'start pecking' (Figure 6.1). (The pecking of the spot is the chick's signal method of asking for food: if the spot was on its parent's, rather than a model bill, the chick would receive a meal.) It is difficult to believe that every time Tinbergen presented a model bill to a chick his activity coincided with a third, unobserved variable that was really signalling to the chick to start pecking; that would however be possible for the natural observation that chicks peck when their parent arrives. Scientists therefore require experimental evidence to test between causes and correlations. Now that we have provisionally fixed what a signal is, and how one may be recognized, let us consider some examples of signals — the songs of birds, the pheromones of moths and ants, and the dance of honeybees — before we consider the theoretical question of why signals have evolved in the form that we see in nature.

6.2 Bird song

Birds, like mammals, produce sound by blowing air from their lungs over vibratory vocal chords in the trachea, although in birds the vocal cords are situated slightly closer to the lungs than in mammals. Bird song is a very familiar kind of animal behaviour. It has been celebrated by poets, and enjoyed by most people. It is also particularly easy to study, because sound can be recorded and reproduced by a tape-recorder. So, why do birds sing? And why do they sing in the way they do?

Male birds do most of the singing, which gives an immediate clue that singing has something to do with sex. In fact singing seems to serve two main functions in birds: defending territory, and attracting and stimulating females to mate. The following experiment shows the importance of song in territorial defence. John Krebs removed the resident pairs of great tits (*Parus major*) from their territories in Wytham Woods near Oxford. Some of the territories he left empty; but in others he placed loudspeakers, broadcasting the song of a great tit. He then watched to see how long it took another pair to occupy the two kinds of territory. It took longer for the territories with loud speakers to be occupied than it did for the silent territories (Figure 6.2). He then took the experiment a stage further. Great tits sing a repertory of one to eight distinct songs. Why do males sing so many songs? What can eight songs do that one cannot? As before, Krebs removed pairs from their territories and put loudspeakers in instead. This time the loudspeakers in some territories broadcast one song, whereas the loudspeaker in other territories broadcast a

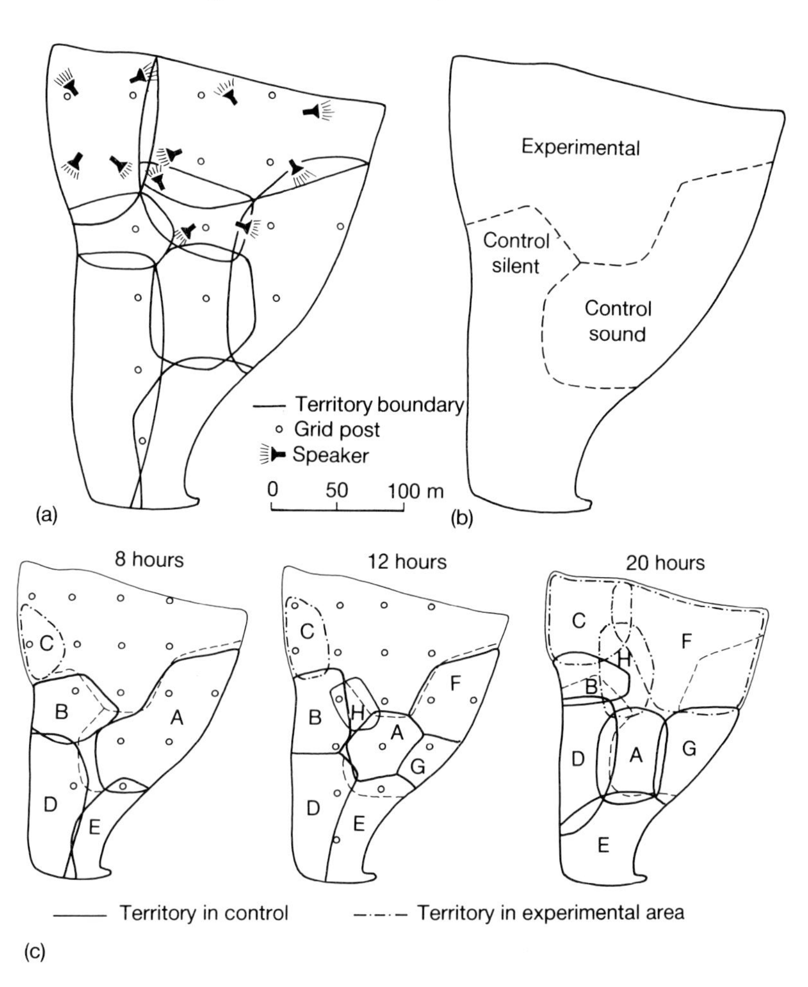

Figure 6.2 An experiment shows that song in great tits functions in territory defence. (a) Territories' boundaries before the experiment. (b) All territorial male great tits were removed from the area, which was then divided into three regions: in the experimental area loudspeakers broadcast the songs of male great tits and two control areas were either silent, or had loudspeakers broadcasting the tune of a great tit played on a tin whistle. (c) The course of re-occupation by intruding great tits: the experimental area was occupied more slowly, presumably because the song deterred intruders. (After Krebs)

repertory of up to eight songs. It turned out that the territories in which a larger repertory was broadcast took longer to be re-occupied. For instance, it took 14 more hours for a territory with eight songs to be re-occupied than one with only one song. In the great tit, then, males sing to keep intruders out of their territories, and a repertory of songs is a more effective deterrent than a single song.

Male birds may also sing to stimulate females. The American ethologist

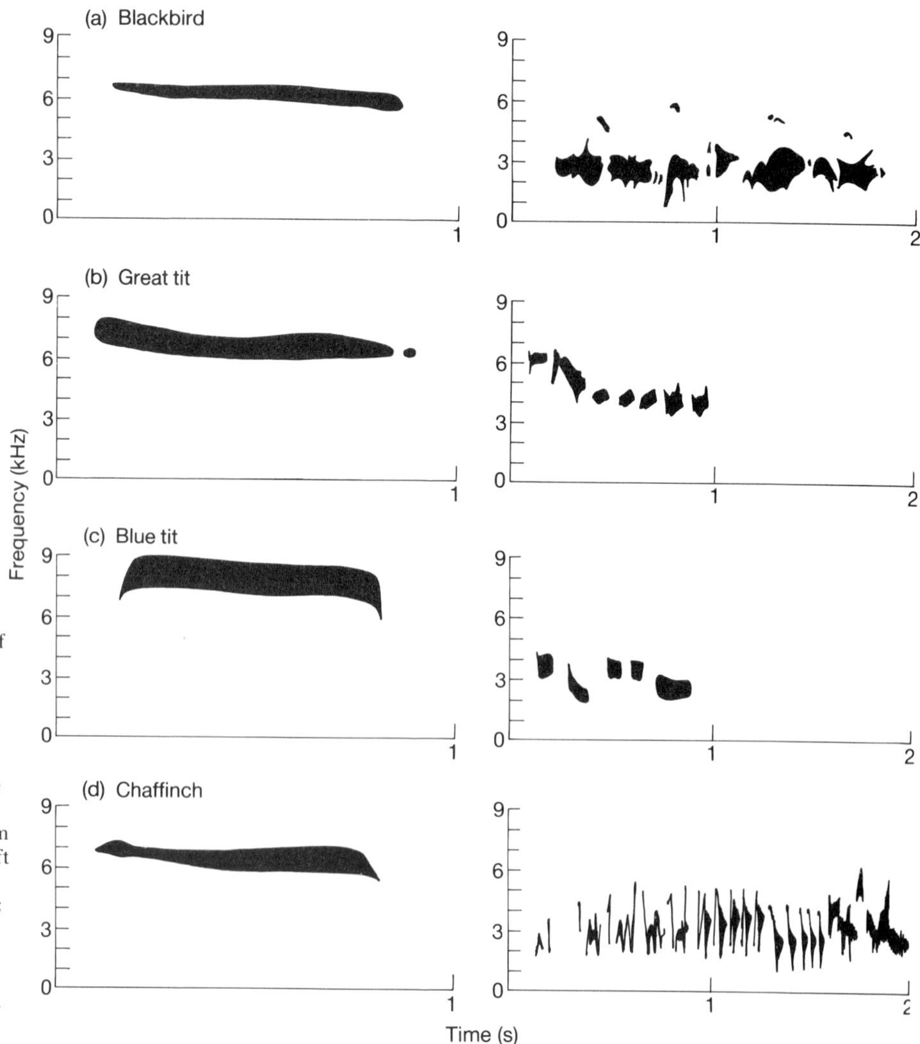

Figure 6.3 Many species of birds give alarm calls when they see a dangerous predator. The particular bird giving the alarm call runs the risk of attracting the predator to itself, so the alarm calls have acoustic properties which make them difficult to locate. On the left are the alarm calls of four species of song bird species; their similarity is suggested by comparison with the ordinary song of the four species, shown on the right. (After Marler)

Don Kroodsma did an experiment in which he played normal and artificially reduced repertories of male canary song to female canaries. The females who were played the reduced repertories turned out to build nests at a lower rate. Another, less direct, observation which suggests that song serves to stimulate females is that males in polygamous species of North American wren have more complex songs than have closely related monogamous species. Competition among males for mates is stronger in polygamous than in monogamous species. If songs have been evolved by males because of competition for mates, one would expect males of polygamous species to have more complex songs than monogamous species.

Territory defence and mate attraction are not the only functions of song. Another kind of song is the 'alarm call'. Individuals of many bird species give alarm calls when they spot a dangerous predator. A blackbird (*Turdus merula*), for example, might give an alarm call on seeing a hawk flying overhead. The alarm call stimulates other nearby blackbirds to take evasive action. As was first noticed by Peter Marler, the alarm calls of many species sound similar (Figure 6.3). They all share certain acoustic properties which, he thought, would make the call difficult to locate. Making a noise when a predator is nearby is, after all, a dangerous thing to do; it may attract the predator to the noisy bird. There would be an advantage in giving an alarm call that is difficult to locate. An experiment of Charles Brown has confirmed Marler's hypothesis. Brown measured the accuracy with which horned owls (*Bubo virginianus*) and red-tailed hawks (*Buteo jamaicensis*) oriented to recordings of mobbing calls (which are not acoustically camouflaged) and alarm calls of the American robin (*Turdus migratorius*). He found the predatory birds oriented less accurately to the alarm calls, as Marler would have predicted.

6.3 Pheromones

Pheromones are chemical signals, released by one individual and smelt by another, whose behaviour is thereby influenced. They are smell signals. Humans, as it happens, make little use of smells in communicating with others; but in some other species, such as ants, it is the main method of communication.

The best known pheromone is emitted not by an ant but by the female silk moth (*Bombyx mori*). The pheromone is called bombykol, and attracts male silk moths. We humans cannot smell bombykol, but it is potent in its effect on male silk moths. They can smell the bombykol released by a single female, at a distance of over a mile. The male's olfactory organs are his antennae, which are enlarged and have a fine mesh of side-branches; they are so sensitive that, according to the experiments of Dietrich Schneider, a single molecule of bombykol elicits a response in the sensory neuron of the male. Only about 200 molecules need to hit the antennae of the male, for him to fly off in search of the female. The antennae also show a very specific response: they sense only bombykol, and are not stimulated by even very similar molecules. Once a male silkmoth has sensed bombykol, his only task is to fly in the correct direction; he must make the correct taxic response (p. 40). He could achieve this by either of two techniques. He could fly around, measuring the pheromonal concentrations, and then orient in the direction where the concentration increased; or he could simply turn upwind. In fact, such evidence as there is suggests that male moths follow the second rule: on sensing bombykol they simply fly upwind. If at any point they lose the scent they fly in zig-zags from side to side until they catch it again, and then fly off upwind again. That set of responses is enough to guide them to the female.

Female ants also release pheromones to attract males; their pheromones, however, are only effective over distances of 25–30 yards. The social behaviour of ants is controlled mainly by pheromones: they do use visual and auditory signals, but most of their signalling is by means of smells. Not only mating is controlled pheromonally, so too are finding and exploiting food, recruiting nest mates for battle, warning about enemies. The collection of scent glands employed by an ant such as *Iridomyrmex humilis* (Figure 6.4)

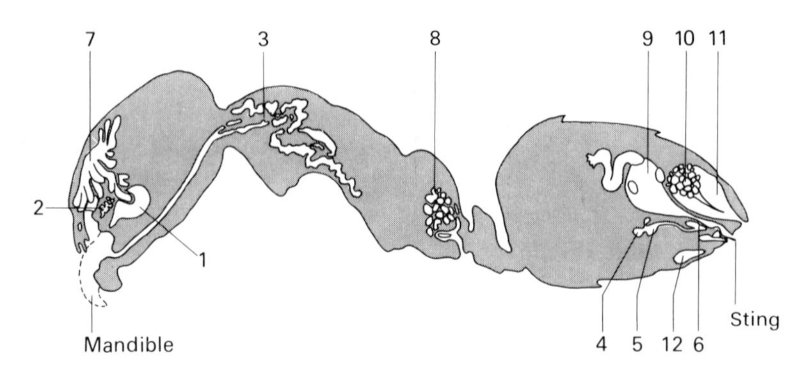

Figure 6.4 Pheromones are manufactured and released from special glands. Here are the 12 main pheromone-releasing organs of the ant species *Iridomyrmex humilis*. 1 mandibular glands, 2 maxillary gland, 3 thorax labial gland, 4 poison gland, 5 vesicle of poison gland, 6 Dufour's gland, 7 postpharyngeal gland, 8 metapleural gland, 9 hind gut, 10 anal gland, 11 reservoir of anal gland, 12 Pavan's gland. (After Wilson)

beats the range of mere human cosmetic collections, and all the ant's scents are meaningful.

Let us consider how several ant species recruit nest mates for group action. If, for instance, a lone ant finds a food source too large for it to bring back to the nest by itself, it runs back to the nest leaving a pheromone trail on the way. Different kinds of ants leave trails from different pheromone-releasing organs. *Solenopsis*, for instance, releases its trail pheromone from its Pavan's gland; *Myrmica* from its poison gland; *Lasius* from its rectal gland. When the food-finder arrives at its nest it uses further pheromones to recruit other ants to come and collect the food; *Myrmica rubra*, for instance, having laid its trail from its poison gland, attracts its nest mates back to the food with a pheromone from its Dufour's gland. Recruitment is not always effected by mass-acting pheromones. The food-finder might instead single out one other nest mate, and then lead it alone back to the food source. They match the recruiting party to the size of the food source. Bert Hölldobler has discussed how *Leptothorax* workers may recruit and lead a single ant when the food source is too big for one ant but only needs two to move it. When, for instance in *Solenopsis*, the food source is so large that many ants are needed, mass-acting pheromones are released.

The scent trail of the ant is released as a liquid (which evaporates) on the ground; but it is known that it is perceived by scent rather than taste. In an experiment, leaf-cutter ants of the species *Atta texana* had to follow a trail by walking along a plastic roadway placed just above the trail; they followed the trail as usual, but must have done so by the airborn odour of evaporated scent. Like the silkmoth, ants sense pheromones through their antennae; but they make continual use of both antennae to keep them in the right direction. Unlike the moths, which, we have seen, ignore relative pheromonal concentrations, and simply fly upwind, ants steer by balancing the pheromonal concentration to the right and left. If the concentration increases to the right they turn in that direction until both antennae are sensing equal concentrations: and the rule will guide them straight down the odour trail. In an experiment, Hangartner placed two trails in parallel; to begin with they were of equal concentration, but as they proceeded one of the trails became progressively weaker. The *Lasius fuliginosus* in the experiment started by walking between the two trails, and then moved across to the stronger trail in such a way as to balance the odour strengths sensed by their two antennae (Figure 6.5).

Recruitment pheromones are not confined to the exploitation of food: they

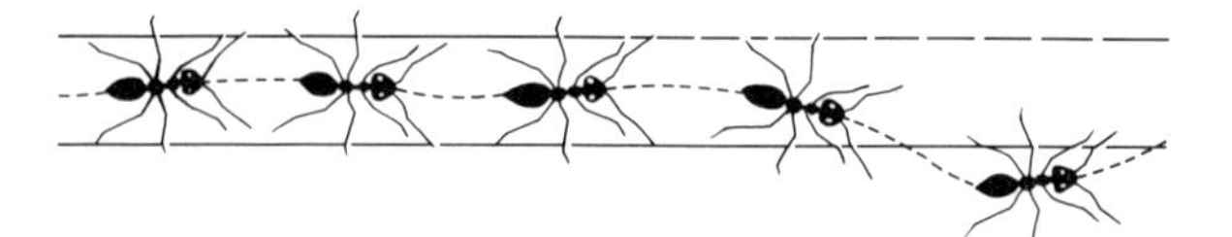

Figure 6.5 Ants follow pheromone trails by balancing the concentrations smelled through its two antennae. These *Lasius fuliginosus* are walking along two experimental trails: the bottom trail has constant concentration; the concentration of the top trail starts the same as the bottom one but gradually decreases. The ants initially walk between the two but cross over to the bottom one at the point predicted if the ants are balancing the concentrations measured on either side.

are also used to recruit armies for territorial disputes, or (in certain species) for slave raids. Not all ant species enslave other ants; but *Formica subintegra* (for example) does. It raids the pupae and larvae from the nests of other ant species, brings them back to its nest, and when the pupae and larvae later hatch they work as slaves for their captors. These slave raids are co-ordinated by a pheromone from the ant's Dufour's gland (which is very enlarged in this species Figure 6.6). E. O. Wilson calls this pheromone a 'propaganda substance'.

With this kind of warfare going on between nests, the ants need to be able to distinguish their own nestmates from members of other nests. This distinction is made possible by other phermones, called colony odours. Each nest has its own distinctive smell.

Ants warn their nestmates about enemies with alarm pheromones. On scenting an alarm pheromone, an ant may do any of a variety of things: it may run away from the source of the scent; it may freeze and 'play dead', or it may

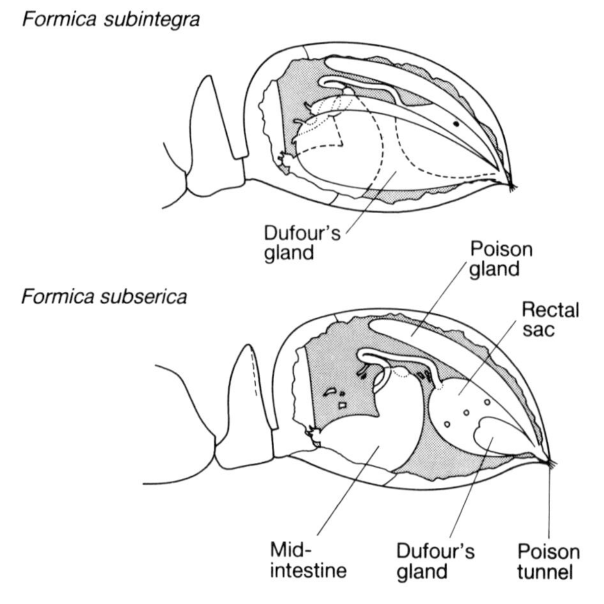

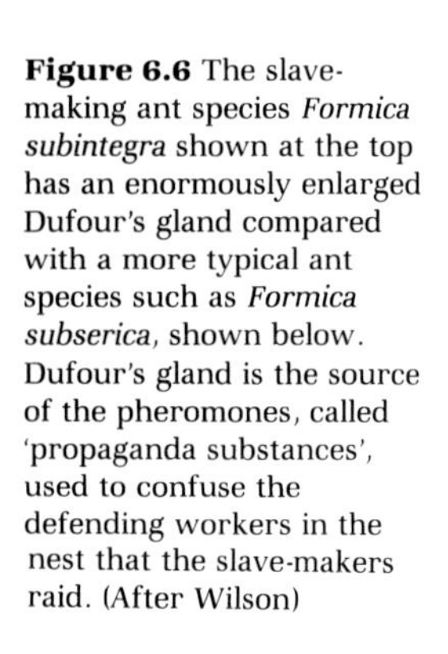

Figure 6.6 The slave-making ant species *Formica subintegra* shown at the top has an enormously enlarged Dufour's gland compared with a more typical ant species such as *Formica subserica*, shown below. Dufour's gland is the source of the pheromones, called 'propaganda substances', used to confuse the defending workers in the nest that the slave-makers raid. (After Wilson)

run towards the source of the scent and attack any nearby enemies. The different responses are probably stimulated by different alarm pheromones. The diversity of messages in an alarm pheromone is suggested by a chemical analysis carried out on the weaver ant *Oecophylla longinoda*. Its alarm pheromone contains over thirty different chemicals. The effects of only four of these have been tested. One makes the recipient ant generally 'alarmed', another makes it run towards the source, and two others make it bite. The alarm pheromone of the weaver ant is a complex message stimulating a whole series of responses in its nestmates.

6.4 The dance of the honeybees

6.4.1 Its meaning

If you put out a dish of syrup during the summer, it might be days until the first bee found it. Bees treat a new dish of syrup like a new flower, recently opened and full of nectar: its fortunate discoverer feeds on the syrup, and then flies home to her hive. There will now be a much shorter time delay until the next bee comes and during the next few hours lots of bees, all from the same hive as the original discoverer, will visit the food source. The reason for this sudden rush of bees is that the first bee tells her hive mates about the syrup. She tells them where it is, how far, and in what direction; she tells them what the food tastes and smells like, and she also tells them how good a food source it is. All this is disclosed in a special 'dance', the discovery and translation of which is one of the great achievements of modern ethology. It was worked out by the Austrian ethologist Karl von Frisch in the middle of this century, by methods we shall discuss in the next section. But how, first, does the dance work? Inside the hive, where the dance is performed, it is dark and the bees are moving around on the vertical wall of the honeycomb. The bee that has found a food source does one of two main kinds of dance according to how far away the food source is. If it is near the nest she does a 'round' dance, in which she walks round in circles, reversing the direction every turn or two (Figure 6.7a). The round dance does not tell the recruits the direction to food, but it does tell them that food is nearby. If the food is further away, the bee does a 'waggle' dance (Figure 6.7b). The distance from the food source to the hive at which the changeover from round to waggle dance takes place differs between different races of bees; *Apis mellifera lamarcki*, for example,

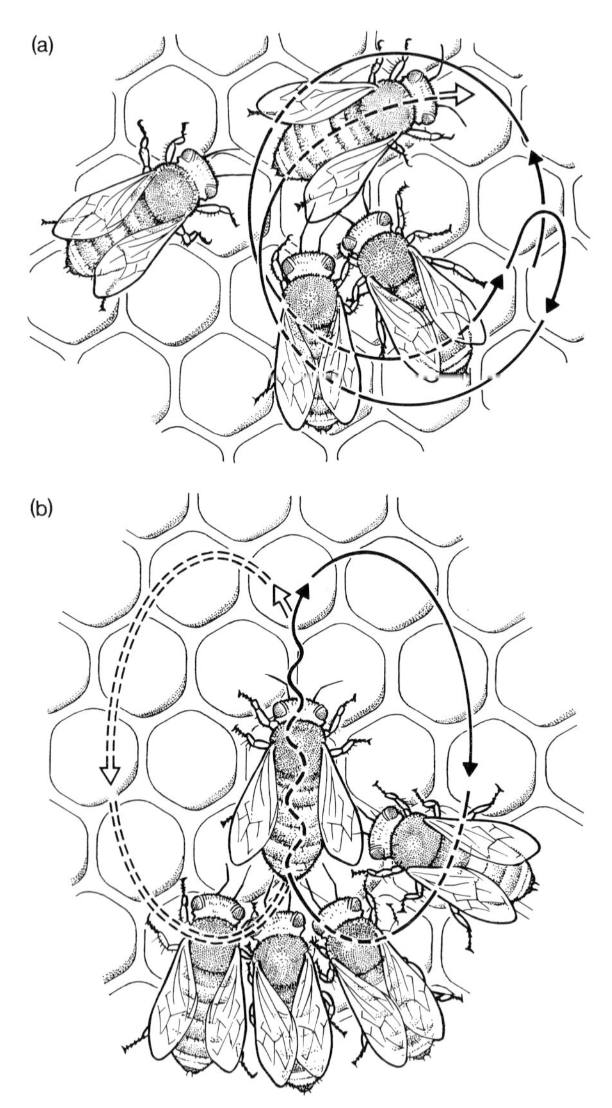

Figure 6.7 The dance of the honeybee tells the location of food sources in relation to the hive. (a) The 'round' dance, which says that food is nearby, but does not show the direction to it. It is used when the food source is so close that the recruits will have no difficulty in finding it without being told the direction. The dance consists of a series of clockwise and anti-clockwise circles; the top bee in the picture is performing the dance. (b) The waggle dance, which says both the distance to and direction of a food source. The distance is told by the rate of the dance; the direction by the angle of the centre part of the dance, during which the dancer wags her abdomen: the angle of this part of the dance relative to the vertical is the angle of the food source relative to the sun.

do round dances up to distances of only about four yards, whereas *Apis mellifera carnica* do it up to about 16 yards. There is no possibility of misunderstanding between dancer and spectator, however, because the changeover distance is constant within any one hive. The waggle dance is a figure of '8', and tells the direction as well as the distance to the food. The key part of the dance is the run up the centre, between the two circles of the '8'.

While running up the centre the dancer waggles her abdomen from side to side. It is the angle of this part of the dance with respect to the vertical that tells the other bees the direction to the food. They understand straight up as meaning the direction of the sun, and they know which direction is straight up because they are sensitive to gravity. The angle of the waggled part of the dance with respect to vertical symbolizes the angle between the food source and the sun. Therefore if the bee orients her waggle 90° to the left of vertical, the food source is 90° to the left of the sun. The bee tells her audience how far it is to the food by another property of the dance, its rate. If she takes longer to complete a circuit, she means the food is further away. Again, the precise relation between the rate of dancing and the distance to the food differs between races. For example, a rate of seven circuits per second means a distance from hive to food of about 400 yards to an *Apis mellifera carnica*, 330 yards to an *Apis mellifera mellifera*, or 260 yards to an *Apis mellifera ligustica*. Scent also matters. The scout bee sometimes marks the food source with her scent. During the dance the other bees can smell her scent and then use this knowledge in finding the food. The dancing bee also regurgitates food to the recruits to illustrate its taste.

Bees can use the dance to point to things other than food. They also dance about possible new nest sites, and about water sources when water is needed to cool down the nest. The dance is a means of indicating distances and directions in the environment outside; it thus enables the nest, in several respects, to exploit that environment more efficiently.

6.4.2 Decoding the dance language

Karl von Frisch (1886–1982) spent his life discovering unsuspected sensory and behavioural skills in animals. We have seen how von Frisch proved von Hess wrong about the sensitivities of fish (p. 43). Soon after that feud, von Frisch made his even more controversial discovery, of the dance language of the honeybee. We have just discussed its form: let us now consider what evidence led von Frisch to the discovery.

In the 1920s, when von Frisch was carrying out similar experiments on honeybees to those he had donc with fish, he noticed that although initially a lone bee came to his food dish, soon afterwards many bees came. He noticed too that the lone bee that first found the food performed a regular sequence of movements on returning to her hive. At the time he thought this 'dance'

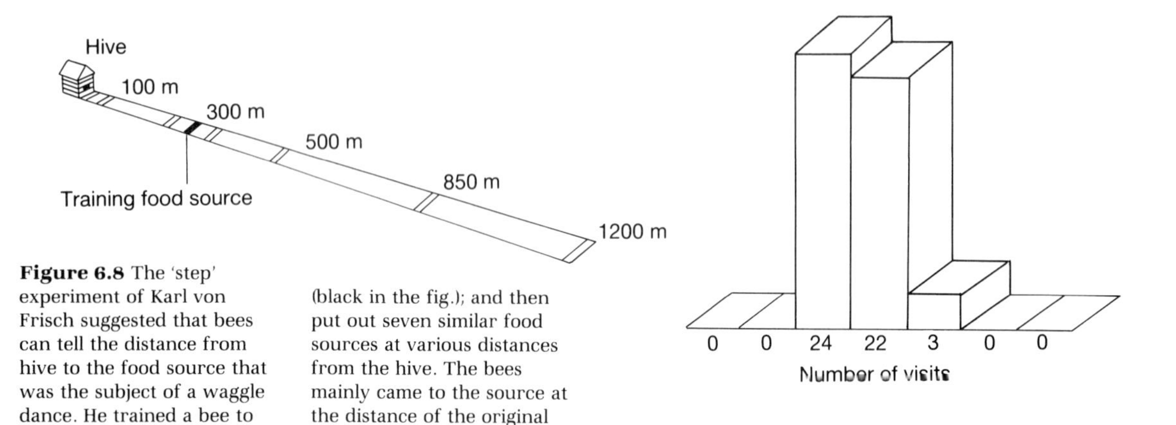

Figure 6.8 The 'step' experiment of Karl von Frisch suggested that bees can tell the distance from hive to the food source that was the subject of a waggle dance. He trained a bee to come to a particular source (black in the fig.); and then put out seven similar food sources at various distances from the hive. The bees mainly came to the source at the distance of the original source.

merely alerted the other bees to the presence of food, which they then located by smell. The bees could learn the smell of the food from the sample regurgitated by the discoverer. Von Frisch did not doubt the odour theory of how honeybees find food until the 1940s. Two kinds of experiment then led him to think again; in their most advanced forms, they are called the 'fan' experiment and the 'step' experiment. In a step experiment (Figure 6.8) von Frisch trained a bee to come to a dish of scented, sugary water. While the bee was back in

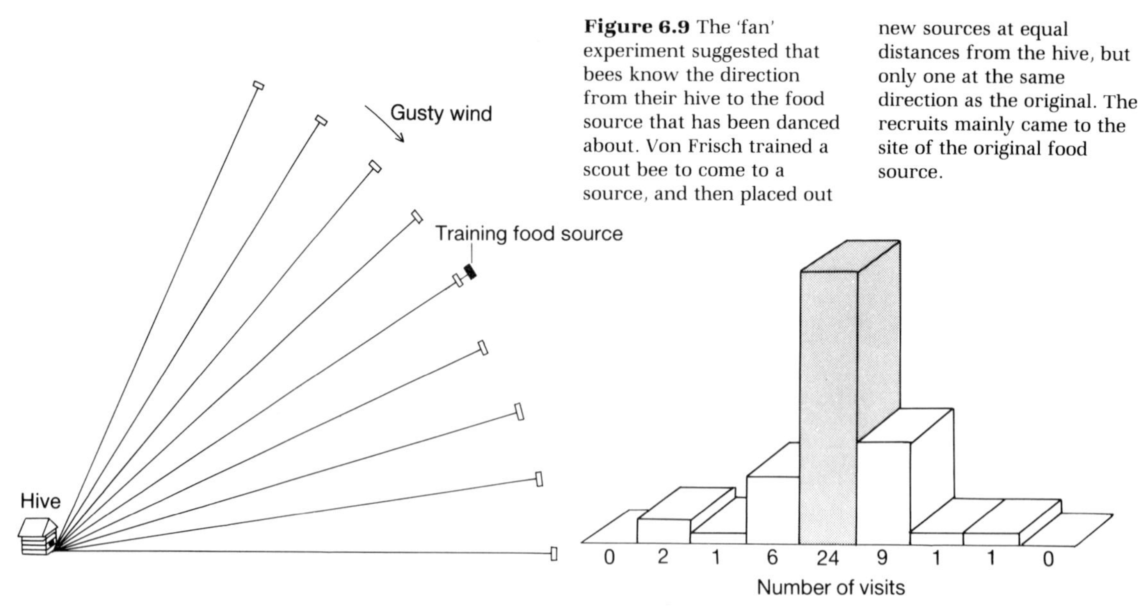

Figure 6.9 The 'fan' experiment suggested that bees know the direction from their hive to the food source that has been danced about. Von Frisch trained a scout bee to come to a source, and then placed out new sources at equal distances from the hive, but only one at the same direction as the original. The recruits mainly came to the site of the original food source.

its hive dancing about the source, von Frisch put out other dishes containing the same scented sugary water nearer to and further from the hive, but all on a line with the original dish. If the bees found the food by smell they would go to the nearest dish; but, in fact they went to the original dish which the first bee had been to. The idea is similar in a fan experiment (Figure 6.9). In the step experiment, other dishes are put in line with the original dish. In a fan experiment, other dishes are put at the same distance as the original dish, but in different directions. Again, all the dishes contained the same scented food, but the bees mainly went to the dish that the first bee had been to. Meanwhile, von Frisch had looked at the honeybee dance more closely. He noticed that he could predict the direction to the food source form the dance. As he varied the position of the food source, the dance varied in a predictable way. He concluded that the bees signalled the direction to the food by the dance.

Von Frisch's theory was initially not believed. However, through the 1950s and early 1960s it became widely accepted by biologists. It was not seriously challenged until the late 1960s, when A. Wenner and others pointed out that von Frisch had not excluded the possibility that the returning, dancing bee brings back with it a smell that is specific to the locality where the food is. A series of inconclusive experiments failed to settle the issue until in the 1970s J. L. Gould finally confirmed von Frisch's theory. Gould made use of the fact that in the dark the bees orient their dance to gravity but if they can see the sun they dance directly with respect to it. They will also dance to an electric light bulb as if it were the sun. In his experiment, Gould painted over the ocelli (small light-sensitive organs) of some of the members of a hive, which would then dance as if it were dark, but otherwise behave normally. He trained the bees whose ocelli had been painted to a food dish, and then let them dance to their unpainted hive mates under an electric light bulb. The dancing bees danced with respect to gravity; the spectators, whose ocelli were unpainted, interpreted the dance as being with respect to the light bulb. If von Frisch's critics were right, and the dance was irrelevant, the bees should find the food as usual. If, however, the dance told the bees where to go, they should go off to the wrong place (depending on where Gould had put the light bulb). They went to the wrong place. Gould, moreover, could exactly control where the bees would go to by moving the light bulb. His experiments show that the bees must be taking notice of the dance, and have confirmed von Frisch's theory more strongly than ever.

6.5 The evolution of signals

Why do animal signals have the properties that they do? One influence is the environment in which the signal is naturally broadcast. Different kinds of signals work better in different environments. Signals that use reflected light, for instance, cannot be used in the dark. Likewise, light does not penetrate through water as well as it does through air, whereas with sound and chemicals it is the other way round. Animals communicating in the dark or through water therefore are more likely to use sound or chemical signals; and terrestrial, diurnal animals make more use of light signals. A second influence is ancestry. Signals usually resemble other activities, performed for other reasons, in the behavioural repertory of an animal. Signals are thought to have evolved by the gradual modification of the activities which they resemble, as they were performed by the animals' successive ancestors.

Three particular classes of activities have been identified as the most common source of signals: intention movements, displacement activities, and activities controlled by the autonomic nervous system. Intention movements are activities that tend to precede some other activity, which is presumably why they often evolve into signals. A familiar human example is the habit of sitting up, or looking towards the door, when you wish to leave the room. The 'upright' signal in the gull is probably derived from an intention movement. 'Upright' is a threat signal, in which the gull stands upright and points its bill downwards and towards the other gull it is threatening (Figure 6.10). This is very similar to the posture a gull physically must take up just before attacking another gull by jabbing it with its bill or biting it. Ancestral gulls therefore

Figure 6.10 The aggressive 'upright' display of gulls is illustrated here by a black-headed gull. (After Tinbergen)

probably learned to fly away from gulls pointing their bills from an upright posture; the posture would then be acting as a threat signal. During evolution it could have been modified into its present form. It has become a pure threat signal rather than just the first stage of an attack.

The reason why displacement activities and autonomic activities should evolve into signals is less clear. Indeed, the function of displacement activities themselves is uncertain. They are actions which appear irrelevant to the circumstances, and are generally performed at times of motivational conflict. A bird, for example, on seeing a conspecific nearby may be simultaneously inclined both to approach it and to flee from it: a motivational conflict between attack (or courtship) and flight. It may then, of all things, preen its feathers, an activity relevant to none of its options. Now, in the courtship of many species of ducks (see Figure 1.8, p. 16), the male points his bill towards his back feathers in what looks like a stylized modification of preening. For this reason, Tinbergen suggested the courtship signal was ancestrally a displacement activity, which is just the kind of activity one would expect from an animal during courtship. The facts are not in doubt, but the function remains obscure. If displacement activities really are irrelevant, why do animals perform them? Perhaps they are, in some manner still unknown to us, not irrelevant; the judgement of irrelevance is that of the human observer, not the animals concerned. But it can be pointed out that times of motivational conflict are just the times when signalling may be expected. They are times when the behaviour of an animal is uncertain; which is just the kind of time when other animals will want information about what it is going to do. When an animal is predictably going to continue to do what it has been doing, a signal is uninformative.

The third source of signals, autonomic nervous activities, was first noticed by Darwin, although their nervous control was not then known; he instanced shivering at times of fear in humans. The autonomic nervous system was not discovered until after Darwin's work. It is a system of nerves, separate from but connected with the central nervous system, and it controls most of the unconscious activities of our body, together with the responses of arousal, rapid breathing, sweating, and so on, characteristic of the 'fight or flight' reaction. Desmond Morris later drew attention to several signals that resemble activities produced by the autonomic nervous system such as auditory signals (shouts) that may have evolved from rapid automatic breathing, and even the scent-marking by urine at territory boundaries in mammals.

So much for the ancestral sources of signals: let us consider the evolutionary process by which they are modified from ancestral behaviour to elaborate signal. The process is called 'ritualization', and there are two main theories about why it happens. They follow from different assumptions of why animals give signals. The older theory supposes that animals mainly tell the truth when signalling. Ritualization then would take place to make the signal less and less ambiguous, and the signaller a more and more accurate advertiser of its intentions. A more recent theory supposes that animal signals are mainly selfish, that animals give signals in order to manipulate other animals into doing things that benefit the signallers. When a baby cuckoo, for example, screams its equivalent of 'feed me! feed me!' at its foster parents, it is not to their benefit. They only feed the cuckoo because it has successfully manipulated their behaviour. In this case, it would be uncontroversial to call the signal manipulatory; but other signals would be more open to disagreement. Be that as it may, if signals are manipulatory, ritualization takes place because of a kind of evolutionary 'arms race'. When the recipient of the signal sees through the deception, there may be an advantage to the signaller in producing a more extravagant signal. Ritualization will then take place, over evolutionary time, in escalating rounds of ever more extravagant signals, to which in turn the recipients respond for a while before they discover (in the case of the cuckoo's foster parent) the distinction between the begging of cuckoos and that of their own young.

Which of the two is correct? As the examples of bird song and bee dances suggest, the first theory applies to some signals, and the second to others. The elaborate songs of male birds clearly fit best a manipulatory interpretation. Male birds compete for territories and mates, and any change to a song that makes them more effective in competition will be favoured by natural selection. There is no reason why a male bird should signal accurate information to its competitors, except in so far as it will scare them away. Likewise, if a male bird sings to court a mate, natural selection will only adjust the song to increase its powers of courtship; the unambiguous informing of the female will be only incidentally relevant.

The signals of social insects, such as the pheromones of ants and the dance of the honeybee, are probably honest rather than manipulatory. For reasons that we shall discuss in Chapter 9, the members of a colony of social insects co-operate with one another, and they will therefore give signal that are as informative as possible; they will not usually give deceptive signals to fellow members of their colony. In many other cases it may be difficult to decide whether a signal is or is not deceptive; but if the signal concerns something over which the animals are competing, it may evolve by an arms race of deception and the discovery of deception; but if it concerns something over which the animals are co-operating, it may evolve to become more accurately informative. The reason why many ethologists would now prefer to explain ritualization as an arms race is that they have increasingly come to realize that selfish competition among individuals is the rule in the animal world. Co-operative behaviour may evolve (see Chapter 9) but even then it will be strictly limited and always liable to selfish exploitation. Therefore, although there will be examples of honest signals, manipulatory signals are probably commoner in nature.

6.6 Summary

A signal by one animal indirectly leads another animal to change its behaviour. Signals are mainly detected by observing behaviour to see which activities predictably lead to changes in the behaviour of others, but this kind of evidence is philosophically unconvincing, and is best supplemented by experiment. The form of signals can be understood in terms of the medium in which they are transmitted, their ancestry, and the process called ritualization by which they evolve from their ancestral form into signals. Ritualization may take place in order to make a signal more, or less, honest according to the type of signal. If it concerns a resource over which the signaller and recipient are competing, the signal will evolve in the direction of deception; if the resource is exploited co-operatively, the signal will evolve in the direction of accuracy. Male great tits sing in order to warn other males away, as can be shown by leaving a loudspeaker playing great tit song in an unoccupied territory; more complex songs are more effective warnings. Male canaries sing at least in part to make females prepare for reproduction. Many birds give alarm calls to warn of danger, but it is not certain whether the acoustic properties of the alarm calls make them difficult to locate. Chemical signals are called pheromones. The female silkmoth uses a pheromone called bombykol to summon males; ants use pheromones for many purposes — recruitment, propaganda, warning of danger. Honeybees tell hivemates the direction, distance to, and nature of, new food sources by special dances. Von Frisch initially thought that the dance merely alerted the bees to the presence of a new source, which they found by its scent. 'Fan' and 'step' experiments led him to doubt this, and, combined with close observation of the dance led to the decoding of the language. The experiment, however, did not control against 'local' odours; and further experiments were needed to test whether bees used the dance or local odours to find the food source. These experiments confirmed that the dance language is used.

6.7 Further reading

Marler (1959) and Cullen (1972) and the authors in volume 2 of the series edited by Halliday and Slater (1983) discuss the principles of communication; and Krebs and Dawkins (1984) whether animal signals are honest. The authors in Kroodsma and Miller (1982) comprehensively review bird songs; and see Krebs and Kroodsma (1980) on repertories. Schneider (1974) explains how male silk moth pheromones are received; and Wilson (1971) gives an account of the use of pheromones by ants. Von Frisch (1967) is the classic account of the dance language of the honeybee; but does not cover the recent controversy: Gould (1976) is the best source for that.

7 / Fighting

In nature, many animals are competing for limited resources. Animals could lead more comfortable lives if there were not so many competitors, each seeking more space to live in, more food, and (if they are males) more mates. Animals passively compete for these resources by taking as much as possible for themselves; but they will also, in some circumstances, actively fight for them (Figure 7.1). Fighting is the most overt, naked form of competition for resources.

We might expect that aggressive fighting would be common in nature, because natural selection will favour the most successful animals in competition, and the strongest animals are probably the most successful in fights. Darwin certainly thought that aggressive fighting was common. His work *The Descent of Man* contains several sections on 'the law of battle' in different animal groups; he collected a large number of anecdotal observations, all of

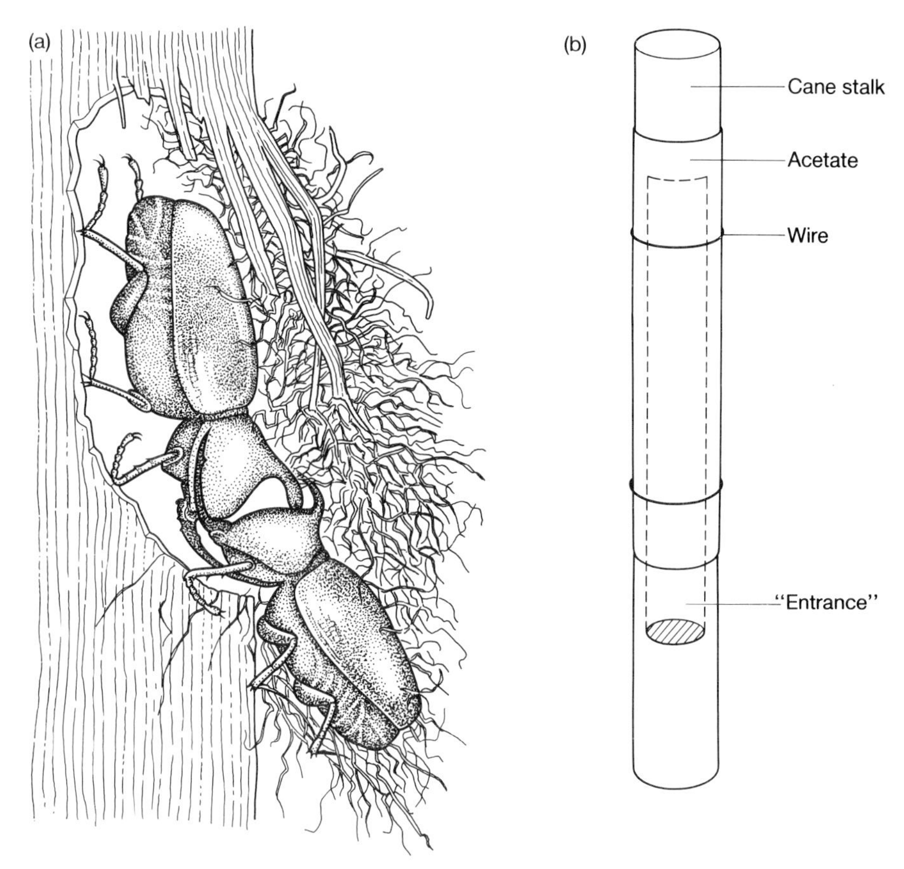

Figure 7.1 Fights among males of the central American lamellicorn beetle *Podischnus agenor* usually take place in vertical tunnels within sugar cane. William Eberhard built artificial cane burrows (b) to watch fights. Fights take place when a male enters a tunnel where another male is already resident. The aim of the fight is to push the opponent down the tunnel to its entrance, clamp him, swing him free, and then let him drop. The resident here has clamped his opponent and is lifting him away from the tunnel entrance. The invader is hanging onto the shredded sugar cane to try to prevent his fall. (After Eberhard)

active fighting, blood-letting and carnage. He does not hint at any restraint in animal fighting. This must be because he did not know any observations of it: Darwin was interested by every possible form of moral behaviour in animals, and if he had known of restrained fighting, he would have written about it.

The more detailed observations of animal behaviour made in this century suggest a different conclusion. Active aggression is now thought to be exceptional. Animals apparently avoid unrestrained battles, and in this chapter we shall discuss three kinds of social behaviour that reduce the amount of aggressive fighting. These facts, however, in their turn pose something of a theoretical paradox. Darwin's own account of the unrestrained 'law of battle' in animals readily fits in with the theory of natural selection; therefore, as his account of the law of battle is now thought to be wrong, we are left with the problem of reconciling new observation and established theory. We shall have to reconsider how Darwin's theory should be applied to animal fighting, to see whether natural selection might in some circumstances favour animals that avoid aggressive conflict.

7.1 Ritualized fighting

Most animal fights are restrained, or, as it was termed by early ethologists, 'ritualized'. Animals avoid using their most powerful weaponry when fighting other members of their species. Rattlesnakes are a clear example. A rattlesnake possesses a powerful poison which it uses against prey and dangerous enemies. However, when fighting against another rattlesnake it does not use its poison fangs. Instead the two rattlesnakes fight in a gentlemanly, if energetic, joust in which each tries to push the other to the ground. The loser, after being floored, retreats. The contestants come out of the fight relatively uninjured.

A ritualized fight will often proceed through several stages, like a tournament. According to an account by Lorenz, fights between cichlid fish of the species *Cichlasoma biocellatum* pass through up to three stages, at any of which one of the contestants may drop out. They start with broadside displays, move on to tail beating, and finish with harmless mouth wrestling. Each fish, according to Lorenz, moves on to the next stage only when the other is ready. The ritualized nature of the contest is made particularly clear by what happens if one of the fish finds itself with a temporary, but irregular, advantage. As Lorenz wrote, 'one of them may be inclined to go on to mouth-pulling a few seconds before the other one. He now turns from his broadside position and thrusts with open jaws at his rival who, however, continues his broadside threatening, so that his unprotected flank is presented to the teeth of his enemy. But the aggressor never takes advantage of this; he always stops his thrust before his teeth have touched the skin of his adversary'. The fighting of male red deer, which has been studied by Tim Clutton-Brock and his colleagues on the island of Rhum off the west coast of Scotland, likewise passes through up to three main stages (Figure 7.2). They start by roaring at each other. The second stage is a kind of broadside display; it is called the 'parallel walk', and in it the two males walk back and forth along side each other. Only after that may they lock antlers and push each other. Red deer do actually inflict physical injuries in fighting; minor cuts and bruises are common, as some 25% of males are so injured each year, but serious injuries such as a

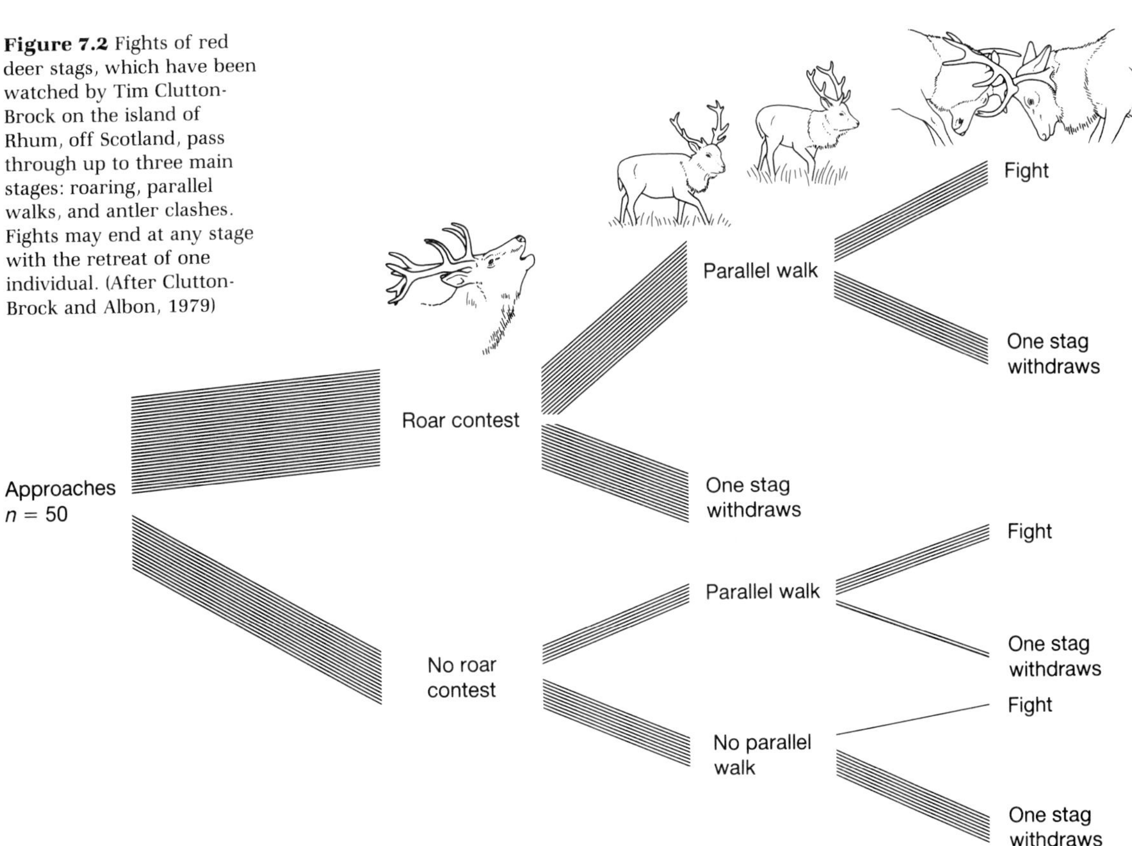

Figure 7.2 Fights of red deer stags, which have been watched by Tim Clutton-Brock on the island of Rhum, off Scotland, pass through up to three main stages: roaring, parallel walks, and antler clashes. Fights may end at any stage with the retreat of one individual. (After Clutton-Brock and Albon, 1979)

poked-out eye, have also been recorded. Six per cent per year of the males watched on Rhum suffered permanent injury. Although the peaceability of animals should not be exaggerated, even in red deer the level of aggression is considerably below the maximum possible. Less than a quarter of contests ever reach the final, antler pushing stage, and even then the deer stag only pushes its opponent's antlers rather than trying to spike his softer flanks; antlers could perhaps be used more dangerously than they are.

Why do animals perform their restrained ritual contests rather than fighting all out straight away? John Maynard Smith has suggested the following subtle answer. An animal that fought dangerously might well beat its gentlemanly opponents. Natural selection would then favour the animals that fought dangerously, which would increase in numbers. But as the dangerous fighters increased in numbers they would increasingly have to fight against other dangerous fighters rather than against timid, restrained fighters. They would then be likely to injure each other badly. The risk of injury might become so great that the timid fighters, which never get hurt because they always run away first, might then be at an advantage. Maynard Smith's explanation, then, is that in nature there is an equilibrium between the risks of injury if dangerous fighting becomes too common, and the advantages of winning fights if dangerous fighting is rare.

The argument is sufficiently clear that it can be formulated mathematically, and the equilibrial balance calculated. In its simplest form, the model requires only two 'strategies', which Maynard Smith called 'hawk' and 'dove'. Hawks are unrestrained aggressive fighters; doves are timid, and run away when threatened. In these terms, the original puzzle becomes that of why natural selection does not produce a population consisting entirely of hawks. Let us suppose that the animals contest a resource which is worth W points to the winner, and zero to the loser. Let us also suppose that serious injuries inflicted by hawks cost D points. (The points have a meaning. They are equal to the effect of a contest on the number of offspring left by the contestants: the model is concerned with what kind of behaviour natural selection will favour.) It is now straightforward to calculate how many points each strategy will recover from each kind of possible encounter. If a hawk meets a dove, the hawk wins W points and the dove takes zero. If a dove meets a dove, neither will fight; but we might suppose that they settle the contest amicably in such a way that each has an equal chance of winning. Then, on average, if an individual fights many contests in a lifetime, a dove wins $W/2$ from its fights

Figure 7.3 Pay-off matrix for the hawk – dove game. The pay-offs are written for contestant A.

		Contestant B	
		Hawk	Dove
Contestant A	Hawk	$\frac{W-D}{2}$	W
	Dove	0	$\frac{W}{2}$

with other doves. And when a hawk meets a hawk? They will have an unrestrained fight; let us suppose that one will suffer serious injury first and be the loser, and that on average a hawk has an equal chance of winning or losing. Then, from many encounters, the hawk's average pay-off will be $(W-D)/2$. The four pay-offs can be written in a pay-off matrix (Figure 7.3). It is easy to see, in the matrix, when one strategy is favoured. In a population mainly consisting of doves, the animals will nearly always play against doves, and 'hawk' will be favoured because its average pay-off is higher in these fights: $W > W/2$. But now consider a population mainly consisting of hawks. Dove wins on average zero, and hawk $(W-D)/2$. Which strategy is favoured depends on the relative values of W and D: if $W > D$ hawk is favoured; if $W < D$, dove is favoured. We now have an almost complete understanding of the model. If $W > D$ then hawk is always the best strategy; but if $W < D$ hawk is best when dove is common, but dove is best when hawk is common. Therefore, the condition for that balance of restraint from which we started out is that $W < D$, that is the value of winning the resource must be less than the cost of injury.

It is not difficult to calculate the exact proportions of hawk and dove that give an equilibrium, if $W < D$. The calculation proceeds as follows. First, for algebraic convenience, let us call the four pay-offs, a, b, c and d (see Figure 7.3) where $(W - D)/2 = a$, $W = b$, $0 = c$, and $W/2 = d$. We know that both strategies are favoured when rare, therefore there must be some proportion at which they do equally well. We shall call that proportion p hawks: p of the individuals are hawks and $(1-p)$ are doves. The equilibrial proportion is found by realizing that, at that equilibrium, hawk and dove must be doing equally well. That is,

average pay-off to hawk = average pay-off to dove.

Now, when p of the individuals are hawks, an individual receives the pay-off it expects against a hawk p times, and the pay-off it expects against a dove $(1-p)$ times. Therefore,

$$\text{average pay-off to hawk} = pa + (1 - p)b$$
$$\text{average pay-off to dove} = pc + (1 - p)d$$

At equilibrium, these are equal,

$$pa + (1 - p)b = pc\,(1 - p)d,$$

and we merely have to solve the equation for p to discover it. The solution is

$$p = \frac{d - b}{a - b - c + d},$$

which if we substitute for a, b, c and d from the matrix (Figure 7.3) reduce to $p = W/D$. The analysis of the model is now complete, although we might round it off with a baptismal ceremony. Maynard Smith calls the strategy that natural selection produces (that is, the strategy that exists at equilibrium) the 'evolutionarily stable strategy' or ESS. When $W > D$ the ESS is hawk; when $W < D$, it is a proportion W/D $(=p)$ of hawks and $1-(W/D)$ of doves. The ESS of hawks alone is an example of a 'pure' ESS and that containing a proportion of each is an example of a 'mixed' ESS. The mixed ESS could be realized in nature either by a population containing W/D hawk individuals and $1 - (W/D)$ dove individuals, or by a population all of whose members behaved as hawks in W/D of their contests and as doves in $1 - (W/D)$ of them.

We can conclude with three general remarks about the ESS, and one in particular about fighting. First, the sense in which the strategy is evolutionarily stable is that no other strategy (among those considered in the model) can do better than it. A population of animals will evolve to an ESS and then remain there: we should expect to see animals in nature behaving according to an ESS. The particular model of hawks and doves is not, however, supposed to represent the strategies of real animals; it is supposed to contain the essence of the biological problem, and allow us to see what natural selection will favour. The other general point is that the theoretical apparatus used to solve the problem is of much more general applicability. Many biological problems have been considered in terms of strategies, pay-offs and ESS. The method is in general applicable to any situation in which the pay-offs to strategies are 'frequency-dependent'; in this model, for instance, the pay-off to hawk depends on the frequency of hawks in the population. In cruder terms, the theory is applicable whenever the best strategy depends on what the other members of the population are doing. Such situations are probably particularly common in social behaviour.

Although the theory has general applicability, it was first used to tackle the problem of ritualized fighting. The solution it offers is that the pay-off to dangerous fighting will decrease as the habit becomes commoner, which will decrease the amount of dangerous fighting in nature. The frequency-dependence in the advantage of unrestrained fighting is probably not the only reason why natural fighting is restrained. A more obvious, but perhaps more important, reason is that animals differ in their strength, and weaker animals will be selected to avoid fights with stronger adversaries. During a tournament like that of the red deer for instance, the opponents can probably size each other up, to see how strong the other is. If an animal is to avoid fighting with another, stronger animal, it must first test how strong his opponent is. The early stages of the tournament may be restrained so that the animals can try their strengths without risking their lives. The roaring and parallel walks of the red deer may be safe trials of strength.

Natural selection does favour those animals that are most successful in the competition for resources, but that does not mean it favours unrestrained aggression. The most successful animals may avoid dangerous fights in order not to be injured by a stronger, or excessively carefree, opponent. The avoidance of conflicts with stronger opponents is the reason why dominance relations develop — they are the subject of our next section.

7.2 Dominance

Dominance is a common, but not universal, kind of relationship between the members of a group, in which some animals, the dominant ones, have priority over others, the subordinate ones. The priority concerns access to such desirable resources as food, places to sit, and mates. The Norwegian biologist Thorleif Schjelderup-Ebbe made the first important observations on dominance, in the 1920s. He studied the common domestic hen. Since then so many ethologists have studied these birds that more is known about dominance in hens than in any other species.

When an experimenter sets up a new group of hens they first fight among themselves. Gradually the hens learn to recognize each other, thereby learning which are the stronger, and which the weaker, individuals. Each hen then learns to give way to stronger hens than herself; she learns not to get into fights she would probably lose. Dominant hens assert their priority by pecking subordinate hens: pecked subordinate hens move out of the way of the pecker, to allow the dominant hen access to the nesting site, or roosting site she was using, or the food she was about to eat. The exact form of the dominance relations within a group of hens depends on the size of the group. There is a simple linear hierarchy in groups of less than about ten hens. This means, in a group of say ten hens, that the 'boss' hen is dominant to all the other nine hens, the second hen is subordinate to the boss hen but dominates the other eight, and so on down the hierarchy. In larger groups the hierarchy can become more complex. 'Loops' may form, for example, in which one bird *A* dominates another, *B*, which dominates another, *C*, which itself dominates *A*.

It is very desirable to a hen to be dominant. Dominant hens take the pick of the food and the roost sites; dominant males also copulate more with females than do subordinate males. Natural selection must therefore be favouring the dominant animals. What kinds of animals become dominant? In hens the dominant birds are usually larger in size. They also tend to have more of the male sex hormone testosterone in their blood: a hen can even be made to ascend its dominance hierarchy by injecting some of this hormone into her blood. There is evidence, from other kinds of animals, of many other factors which affect dominance. Parasites, for instance, affect dominance, at least in mice. W. J. Freeland set up groups of three mice, the different mice having been injected with different quantities of parasites. The mouse with the least parasites usually became dominant. The dominance hierarchy of hens is also arranged by sex. Males are usually dominant to females. If there are many males and females in the group, the males form a separate hierarchy above that of the females. In most species which form dominance hierarchies, males are dominant to females; but this is not universal. There are some species, such as hyenas and vervet monkeys, in which females are dominant to males.

Dominance is found in many of the species of primates that live in social groups. The dominance hierarchies of primates are often more complex, overlapping networks, rather than the simple ladder of the hen hierarchy. However, the hierarchies of hens and of various primates share many features in common. Like in hens, dominance in primates determines access to food, places to sit, and mates; and dominance is determined by similar factors. One different factor from hens is age. Age has little effect on dominance in hens, but in many primates older animals are often dominant to juveniles.

The main reason why dominance relations reduce the amount of overt aggression within a social group is that weaker individuals come to learn that they are weak, and therefore avoid entering into fights that would be effort-

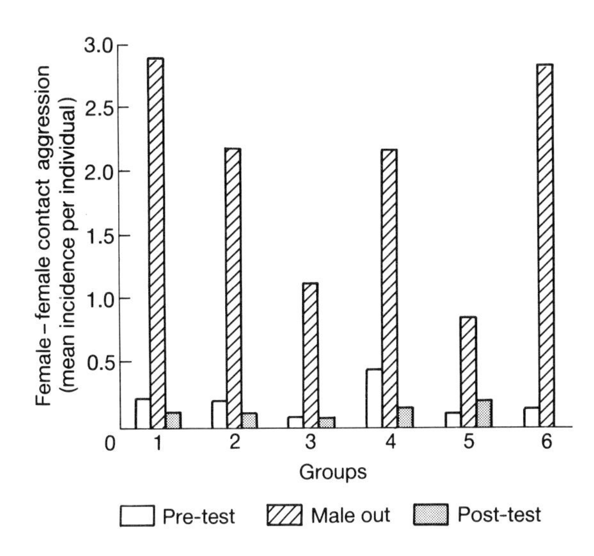

Figure 7.4 In each of six groups of pigtail macaques (*Macaca nemestrina*), when adult males were removed from the group the amount of aggression among the remaining females increased dramatically. When the males were returned the level of fighting went down again. (After Oswald and Erwin)

less to the stronger, and dangerous to themselves. However, additional mechanisms may be at work. In the pigtail monkey (*Macaca nemestrina*) the most dominant male reduces the amount of fighting within his troop by policing any fights that break out. For this reason, when M. Oswald and J. Erwin removed the dominant male from a troop the amount of fighting among the remaining pigtails went up (Figure 7.4); the effect is not simply due to the removal of an individual, because no consistent increase in aggression followed the removal of either a relatively dominant or a low-ranking female. The 'policing' of conflicts, however, is a second-order effect. The establishment of a settled dominance hierarchy itself is the main limit on aggression.

7.3 Territoriality

The three-spined stickleback is a freshwater fish, about 2–4 inches long, which inhabits lakes, ponds and rivers in Europe and North America. It also happens to be a favourite animal for ethological experiments. During the winter the sticklebacks swim around in schools; then, come the spring, with its light, warmth, and improved food, the males start to develop red bellies and become more aggressive. They each start to defend an exclusive area from other sticklebacks. This defended area is the male stickleback's territory. If another stickleback swims into a male's territory, the owner swims hard at the intruder, and drives it off. But he will only attack the other fish if it trespasses onto his territory: if the first fish trespasses onto the other male's territory the roles are reversed. Niko Tinbergen demonstrated the site-dependence of which fish attacks and which one flees in a simple experiment (Figure 7.5). He let two male sticklebacks, *a* and *b*, set up their territories in an aquarium. He then put *a* in one test tube and *b* in another. When he put both test tubes in *b*'s territory, *b* tried to attack *a* through the glass and *a* tried to flee; Tinbergen then moved both test tubes across to *a*'s territory, and now *a* tried to attack, and *b* to flee. A territory, then is an area which its owner tries to defend and from which intruders usually flee. It is a spatially-dependent dominance relationship.

In contests between the territory owner and an intruder, the owner

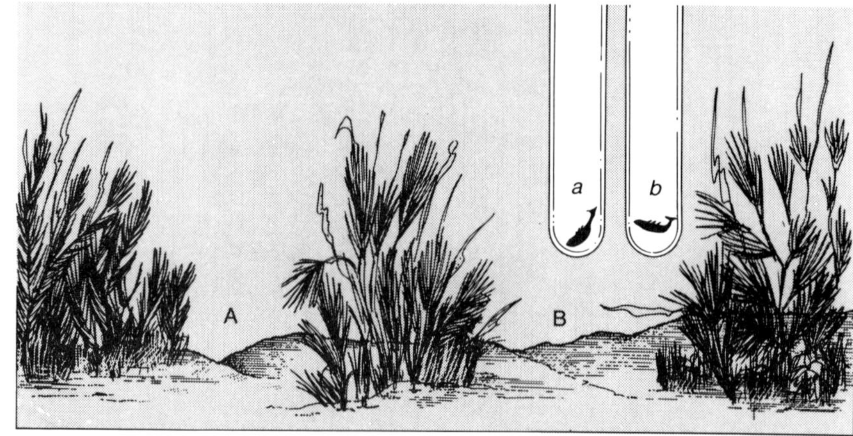

Figure 7.5 Which of a pair of male sticklebacks wins in a territorial dispute depends on where the fight takes place, as a simple experiment by Niko Tinbergen demonstrates. There are two fish, called *a* and *b*, with neighbouring territories. Tinbergen put each fish in a glass tube. When he put the two tubes in *a*'s territory (A), *a* tried to attack and *b* to flee; and vice versa. Who wins depends on place, not relative strength. (After Tinbergen)

usually wins. Why should this be? One reason is that owners are often stronger than intruders, which are animals too weak to have been able to set up a territory. That cannot be the reason in Tinbergen's experiment, however, for which some other explanation must be found. The territory may be worth more to the owner than to the intruder, and the owner, therefore, prepared to fight harder for it. The owner will have spent time finding out about his territory; where the best sites are for hiding, where the best food is to be found. The intruder lacks this knowledge; the territory would, to begin with at least, be worth less to him. A third possible reason is that there may be a convention that 'owner always wins' for settling territorial disputes. The 'owner wins' convention is suggested by Tinbergen's sticklebacks. it is also suggested by an experiment on the speckled wood butterfly (*Pararge aegeria*) by N. B. Davies. The speckled wood defends sunspots underneath the wood canopy. Davies first confirmed that intruders do usually lose contests over territories. He then experimentally removed an owner, and allowed a previous intruder (which had lost against the removed owner) to establish itself in the territory. After only an hour or so, Davies released the ex-owner as an intruder. It lost. The owner cannot have always won because it was stronger: the same two individual butterflies were used in both roles. It seems that a butterfly simply automatically backs down when it meets an owner.

The defence of territories is a widespread habit among many kinds of animals. Different species have probably evolved territoriality for different reasons; but some functions are commoner than others, and two of the main functions are feeding and reproduction. The stickleback's territory is reproductive. Natural selection favours the habit in stickleback males because female sticklebacks will only mate with territorial males. After mating, the territory is advantageous in protecting the eggs and young fish. In other species, territories are defended for energetic, rather than reproductive, reasons. The golden-winged sunbird (*Nectarinia reichenowi*) defends a territory of about 1000–2500 flowers for purposes of feeding; it feeds on the nectar provided by the flowers. F. B. Gill and L. L. Wolf, who studied the sunbird near Lake Naivasha in Kenya, measured the nectar levels in defended and in undefended flowers. There was on average more nectar in the defended flowers. This is because, after a sunbird has sucked out the nectar from a flower, the flower takes some time to replenish its nectaries. A territorial sunbird can time its visits to a particular flower such that its nectar has built up to a high level. Undefended flowers cannot be farmed in this manner, because another bird could come and drink from the flower during the interval before it has restored its nectar to the higher level. Defended flowers can therefore be exploited more efficiently and it can pay a sunbird to defend a territory.

If a bird is defending a territory in order to exploit as efficiently as possible the food resources on it, the bird should ensure that it defends a territory no larger than necessary. It takes energy to defend a territory from intrusive neighbouring birds, and the larger the territory the larger the border that has to be guarded. It might be that if a bird tried to defend too large a territory the amount of energy it would gain from its territorial habit would actually decrease, as it spent more energy on extra defence than it gained from extra food supplies. This idea may be illustrated in a study of the rufous humming-bird (*Selasphorus rufus*) by Lynn Carpenter, D. C. Paton and M. A. Hixon. This humming-bird migrates southwards down the west coast of America during late July and August. The flight burns up energy, and the humming-bird has to stop from time to time during its journey, to defend a territory and

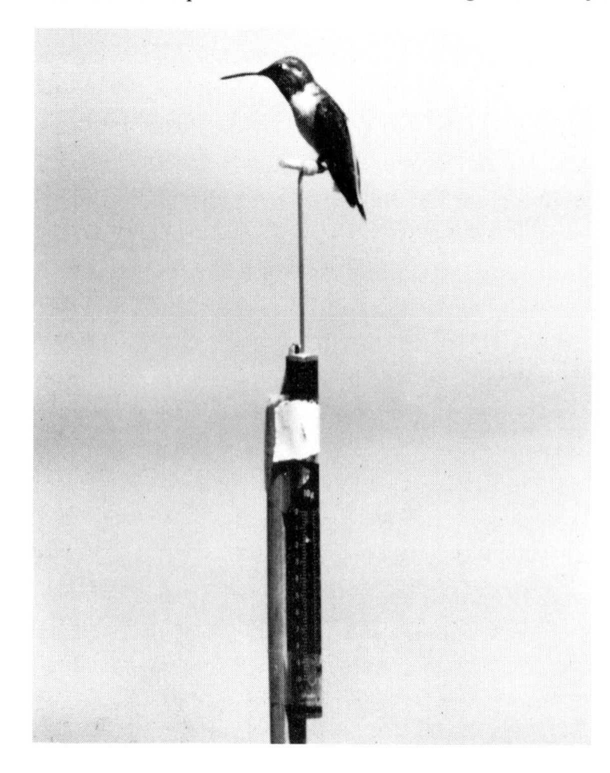

Figure 7.6 A territorial rufous humming-bird weighs itself by perching on a spring balance. (Photo: Mark A. Hixon)

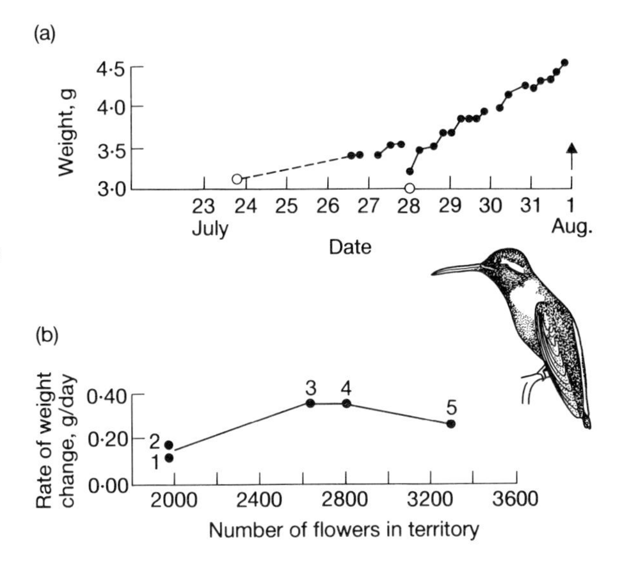

Figure 7.7 (a) An individual rufous humming-bird steadily increases in weight during a week of defending a territory. On 1 August the bird flew on. ○ Weights measured by netting the bird, ● perch-balance weights of free bird on its territory. (b) The weight gain of the birds per day is related to the number of flowers on its territory (territories change size between days). The five points are for five consecutive days, and suggest that the bird has an intermediate optimum territory size. Both graphs are for the same individual; eight birds all gave similar results. (After Carpenter, Paton and Hixon)

re-fuel. Carpenter and her colleagues watched the humming-birds in the mountains of California. There, individual humming-birds whose weight has decreased to 3–3.5 g stop for periods of a week or two; after they have put on about 1.5–2 g weight they then move on again. The rufous humming-bird, like the golden-winged sunbird, feeds on nectar, which it takes from a territory of about 60–4000 flowers. Carpenter kept track of the birds' weight changes by means of perches attached to spring or electronic balances (Figure 7.6). They observed that birds start to put on weight soon after establishing a territory (Figure 7.7a). But it is the relation of the rate of weight gain and the size of the territory that is more relevant here (Figure 7.7b). The rate of weight increase is lower if the territory is too large or too small, relative to an intermediate optimum. The points of Figure 7.7b are for successive days, and it appears that the bird adjusted its territory size to maximize its rate of weight gain.

Humming-birds and sunbirds, then, defend territories in order to feed more efficiently on their defended flowers. The purpose of the stickleback's territory is different. In general, territoriality (like group-living) has different functions in different animals. In all cases, however, the space will be defended to secure access to some limited resources. And despite the universal competition of animals for limited resources, settled territoriality will, like ritualized fighting and dominance, have the consequence of reducing the amount of naked aggression that can be seen in nature.

7.4 Summary

All animals must compete for limited resources, but the competition may not be expressed in unrestrained battle to the death. Nearly all animal fighting is restrained in some way, and contests in some species resemble highly ritualized tournaments. Fighting may be restrained because the advantage of dangerous tactics will decrease as the habit becomes more frequent in the population. If the costs of injury are great, the advantages of aggression will eventually be limited by the risks of injury. The situation can be modelled in an idealized game of 'hawks' and 'doves', which illustrates exactly how natural selection can drive a population to an 'evolutionarily stable strategy' in which fighting is restrained. Animals also restrain their aggression because it is disadvantageous to fight stronger opponents — it is better to run away. Thus 'dominance' relations arise in many social species. The dominance hierarchy may be a simple linear chain, or a complex overlapping network, depending on the size of the group and the species in question. Many factors influence the dominance of an individual — its size, strength, sex, health (in mice, at least), age (in primates, but not in hens). In territorial species, the dominance relations of individuals depend on space. One stickleback will put another to flight when it is an owner attacking an intruder, but will itself flee from the same individual when it is an intruder on the other's territory. The repeated victory of the owner against intruders looks like an arbitrary convention, obeyed by both individuals, that the 'owner always wins'. However, the owners may win because they are more prepared to fight it out than the intruder; when an owner beats a non-territorial intruder, he may win simply because he is stronger. Territoriality, dominance, and ritualized fighting all result in the fact, often noticed, that the amount of naked aggression to be seen in nature is less than the maximum possible.

7.5 Further reading

Richard Dawkins (1976) popularizes the theory of evolutionarily stable strategies; Maynard Smith (1982) provides a more advanced summary. Lorenz (1966) popularized the fact of restrained fighting (together with his own explanation of it, which is no longer accepted). Clutton-Brock *et al.* (1979, 1982) and Clutton-Brock & Albon (1979) describe the contests of red deer on Rhum. Dominance is reviewed by Hinde (1974) and Wilson (1975). Tinbergen

(1953) describes his experiment on sticklebacks; Wilson (1975) again and Davies & Houston (1984) are more recent introductions to the literature on territoriality.

8 / Sexual behaviour

Sexual behaviour poses a set of related questions about mating and the rearing of offspring. Before conception, there may be a struggle among males, and a characteristic (often strange) sequence of behaviour patterns in which the sexes court each other. The term courtship, as used by ethologists, refers to all the behavioural interactions of the male and female which come before, and lead up to, the fertilization of eggs by sperm. Its form and ostentation vary among species. In some species it does not even exist. In others, such as the stickleback (which we shall consider below) it lasts a few minutes. In others it may last for months. Male and female waved albatrosses (*Diomede irrorata*), which live on Isla Española of the Galápagos Islands, may court each other, with an extensive repertory of stereotyped movements of the neck and bill, for several hours a day, day in day out for much of the year (Figure 8.1). The question courtship poses is why it exists, and why, in some species, it has taken on so extravagant a form.

One answer is that courtship ensures that animals mate with other individuals of the correct species, sex and condition. Its extreme development, however, of numerous energetic and colourful displays, and extreme structural modification in males, cannot be so easily explained. It is difficult to believe that all the paraphernalia of sex is needed merely to ensure that males are not confused with females or males of another species: that could be achieved much more quickly and less colourfully. Moreover, the most bizarre sexual characteristics, such as the amazing plumage of male birds of paradise, probably decrease the chances of survival of their bearers. They use up energy, hamper flight, attract predators. They are therefore something of a puzzle. They must possess some hidden function to compensate their obvious disadvantages, because if they did not they would have been eliminated by negative selection. Darwin invented a special theory to solve the problem, his theory of sexual selection. We shall come back to that.

After fertilization one or both of the parents may look after and help to rear the offspring. The question of why parents should look after their young is part of the general question of altruism, which is the subject of the next chapter. But the sexual division of labour — whether it is the male, the female, or both parents who look after the young — is controlled by much the same forces as control other sexual differences and is therefore appropriately treated in this chapter. We shall start by confirming that sexual behaviour does indeed ensure that mates are of the correct species, and then see how well Darwin's theory, modified and tested by recent work, can explain the full variety of sexual behaviour in animals both during courtship and, later, in parental care.

8.1 Choosing a member of the right species

Matings between members of different species are very rare in nature. For example, in a sample of 725 copulating pairs of two species of cicada, which are distinguishable but similar in appearance to the human eye, only seven contained members of two species; the other 718 were matings of the same species. Different species often do not get a chance to interbreed because they live in different geographic areas, or in different localities within an area (one species at the tops of trees, the other on the ground, for example), or they might be active at different times of day. Of the few species that do have the opportunity to interbreed, none do so frequently. They usually do not because the one species does not respond to the sexual lures of the other.

It is adaptive for animals not to mate with members of other species. Hybrid offspring are usually inferior to those produced by matings of members of the same species, often because they are sterile. The best known example of a sterile hybrid is the mule. The mule is the offspring of a he-ass and a mare; if a she-ass mates with a stallion the result is a 'hinney', which is also sterile. Natural selection will act to prevent animals from mating with members of other species if the offspring so produced would be sterile. Natural selection will favour animals that produce normal healthy offspring by choosing to mate with members of their own species, rather than producing sterile hybrid offspring. There have been many investigations of the factors animals use to ensure that they mate with members of the same species. Let us take crickets as an example. There are approximately 3000 species of cricket in the world. They do not all live in the same place, but in any one place there may be more than one, and perhaps half a dozen, species. They can tell each other apart by their distinctive songs. The songs are produced by males when they rub their wings together, and are heard by the female through the ears on her front legs; they are familiar to us from night-time walks, or the vicarious experience of the cinema. The chirrups of the crickets are calling songs, designed to attract females of the same species as the calling male. Females are attracted only by the song, as can be demonstrated by putting out a loudspeaker broadcasting the tape-recording song of a male cricket: females of that species will approach it. Females, moreover, only approach loud-

Figure 8.1 The courtship of the waved albatross, which breeds in the Galapagos Islands, consists of many remarkable behaviour patterns. The pair sway from side to side with necks extended forwards, inspect the insides of each others mouths, click their bills shut, wrestle with their bills, point their heads downwards, and then up to the sky. What is the point of it all? (Photos: Catie Rechten)

speakers playing songs of their own species. They are therefore using the song to choose a mate of the correct species. (Parasitic flies are also attracted to the loudspeaker. Parasitic flies are one of the hazards of the male cricket's sex-life; they are attracted to calling males, on whom they lay their eggs. The growing parasitic larvae soon inactivate the male. It is a common risk in many species that by broadcasting to females, a male attracts predators and other enemies.)

The all-importance of song in the species recognition of crickets can be shown by another experiment. Female crickets can be trained to walk along a 'Y-maze' (Figure 8.2), on which they have to turn to the left or right. If, for instance, the loudspeaker on the left were playing the song of the female's own species, while the one on the right played the song of some other species, the female turns to the left. The side of the songs can be reversed to control for any bias the female may have for turning in one direction or the other.

The Y-maze has also been used to see how hybrid female crickets behave. If a male and female cricket of two closely-related species are forced to mate,

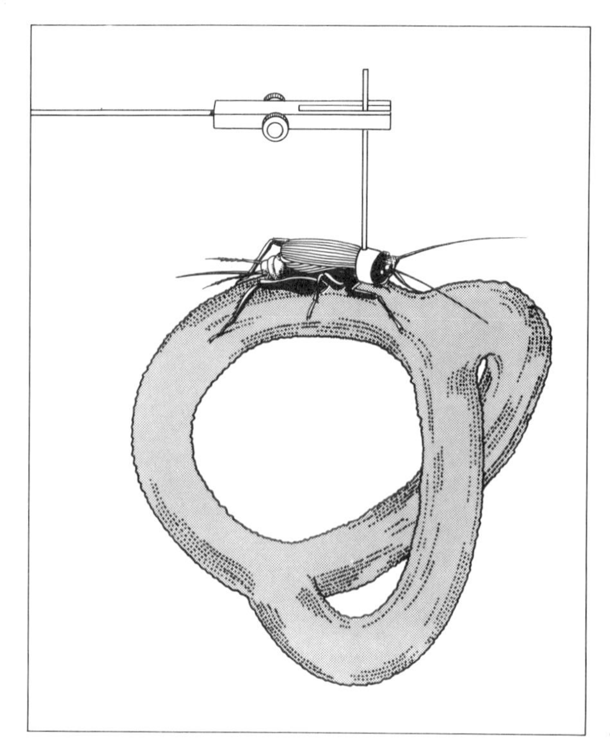

Figure 8.2 As the tethered female cricket walks along the Y-maze it moves beneath her feet. Loudspeakers play the songs of different cricket species from her left and right. Which song she prefers can be measured by which direction she turns on the maze. (After Bentley and Hoy)

some hybrid offspring will be produced. When R. Hoy put these hybrid females on a Y-maze he found that they preferred the songs of hybrid males to those of either parental species. The result is interesting because it bears on the following problem. When a new cricket species evolves, the male song and the female receptor must change in a co-ordinated fashion: a change in one without a change in the other would be disastrous. The fact that the song and the receptor change together in the hybrids suggests that there is some control mechanism which prevents the two from becoming uncoupled from each other; such a mechanism would have the effect of making it more likely that, whenever a male or female cricket changed its song or song-preference during evolution, some other individual of the other sex will have made the complementary change.

8.2 Sexual selection

Darwin put forward his theory of sexual selection to explain the apparently detrimental sexual characters possessed by many species. He suggested that, in those species, males might compete among themselves to determine which males mate with the females, and the females might choose which males of the species they would mate with. If the bizarre characters were favoured by female choice, or in male competition, they could be maintained even if they decreased the viability of their bearers: the advantage in sexual success could compensate the disadvantage in viability. Both male competition and female choice are completely useless from the point of view of the total number of offspring produced: the females would get fertilized at all events. Sexual selection only results in females being mainly fertilized by particular kinds of males.

A modern account of sexual selection would run something like this. We start from the gross difference in size of the male and female gamete. Each member of the next generation is formed by the fusion of one sperm cell with one egg cell; but the sperm cell is tiny, less than one millionth, often less than 100 millionth of the size of the egg cell. A male, therefore, feeding at the same rate and converting energy into gametes at the same rate as a female, can produce gametes at a much higher rate. A male is potentially capable of fertilizing hundreds of females, if he could only copulate with them. A male who did somehow manage to copulate with hundreds of females would be at an enormous advantage compared with a male who only copulated with one or two females. Any adaptation in a male which enables him to copulate with more females will be strongly favoured by natural selection. This is the theoretical basis of male competition. Females, by contrast, gain no advantage by mating with many males. There is no selection on females to copulate with as many males as possible. The rate of reproduction by females is limited by the rate at which they can produce eggs, not by the rate at which they copulate. However, if males differ in their quality, it could pay females to be choosy about which males they will mate with. If one male was defending a better territory than another male, the female might be selected to mate with the better male. This is the theoretical basis of female choice. There is not normally any corresponding selection on males to be choosy about who they mate with. Males are selected to mate with as many females as possible. Any male who chose not to mate with certain kinds of females would mate less than his non-discriminatory competitors.

Natural selection will therefore favour adaptations in males that enable them to mate with more females; but it will favour discrimination in females, if males vary in their quality as mates. This general pattern follows from the cheaper cost of gamete production in males than females. Males can produce sperms at a faster rate than females can produce eggs, and are therefore selected to allocate more of their time to searching and competing for mates than are females. Because males spend less energy in manufacturing each gamete than do females, males can potentially produce more offspring, in a greater number of productive matings. Because males are able to reproduce more, by chance some indeed may do so. The variability in reproductive success of males will therefore be greater than that of females. These sexual differences may be illustrated by an experiment of A. J. Bateman on the fruitfly *Drosophila*. He set up several experimental cages, each containing five males and five virgin females. Only 4% of the females did not mate, and even those were vigorously courted; but 21% of the males did not mate at all, even though they courted hard. The most successful males produced almost twice as many offspring as the most successful females (Figure 8.3). With sexual selection the success of different males varies highly. Some males are very successful, many are not. Females are much less variable in the number of offspring they produce. The greater variability in the reproductive success of males could just be a random result: by chance some individuals will meet more mates than do others, and this will have a larger effect on the variability of reproduction of males, because in males more of the random encounters

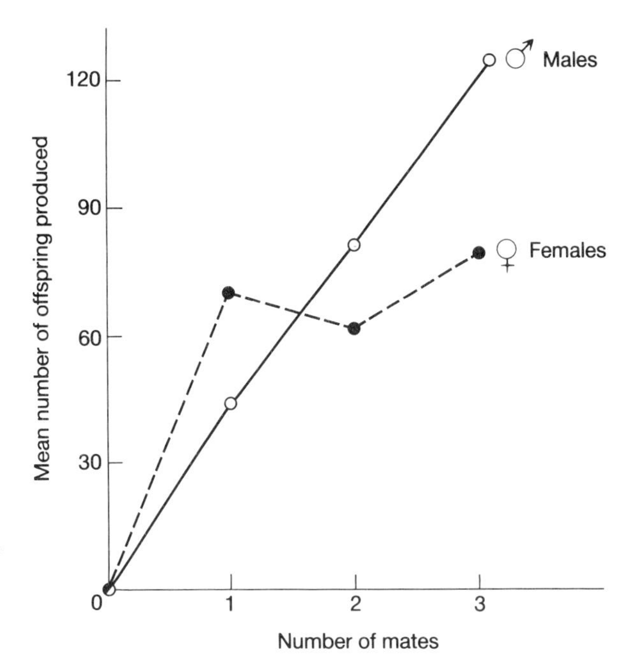

Figure 8.3 The relation between number of matings and reproductive success in experimental fruitfly (*Drosophila*) populations. The number of matings strongly influences the reproductive success of a male, but not of a female. (After Bateman)

can result in mating. If a male meets a female he is unlikely to be prevented from mating merely because his sperms are in short supply (although this can happen); if a female meets a male she may well be short of an egg. This is the fundamental difference from which, according to Darwin's theory, other sexual differences follow.

Chance alone is probably not the only factor influencing the relative variability in reproductive success of males and females. Female choice and male competition will also magnify the difference, for they will increase the variability of male reproductive success. The effect of female choice and male competition will be to make some males even more successful than if those factors were not operating. It would take further work to determine the relative contributions of chance, male competition, and female choice, to the variabilities of reproductive success of male and female fruitflies in Bateman's experiment. But the fundamental theory predicts a strong relation between the amount of energy contributed to the production of the next generation by each sex, which sex chooses mates, and which sex fights or otherwise competes for mates. Let us now consider it in action with some real examples.

8.3 The struggle of males and females

Because the males' reproductive success is probably limited by the number of females they can attract and defend from other males rather than their sperm supply, natural selection will favour any property in a male that enables him to mate with more females. This general principle finds its natural realization in the fascinating diversity of adaptations by which the males of different species seek to increase their share of the mating. Let us examine some examples.

The most obvious form of competition for females is straightforward fighting. We have considered the subject before, in the previous chapter, and we only need notice here the kinds of characteristics it leads to in males. It will favour strength, and fighting in its simplest form will favour increased size. In toads, for instance, the males sit on the backs of females for a few days before the female lays her eggs. Other males try to dislodge the sitting males from the

Figure 8.4 The narwhal's spiral tusk, which is actually an extended tooth (up to 260 cm long), is mainly confined to males. The many functions that have been proposed for it range from sound transmission to drilling holes in ice; but it is really used in aggressive fighting among males. H. B. Silverman and M. J. Dunbar watched narwhals off Baffin Island and repeatedly saw males cross their tusks and strike them against each other. Males also often have body scars and broken tusks, suggestive of combat.

females, by pulling them off. In experiments larger males are much better at dislodging smaller males from females than vice versa. There is an advantage in being large in males, and large males are more likely to mate with females than are small males. In other species males have evolved special weapons with which they fight over females. The narwhal's tusk is an example (Figure 8.4). It is found only in males, who use their tusks to fight each other. Most males suffer severe wounds from these fights and in one sample from a narwhal population over 60% of the males had broken tusks. Other strange weaponry can be seen in the males of some kinds of beetles, and the antlers of deer have evolved for the same reason. In species that fight, a male can increase his effective strength by forming a coalition with another male, and ganging up on competitors. Male lions do just that. Groups of two or three male lions try to take over harems of females by forcibly evicting the existing male owners. Bigger coalitions of males are more successful in taking over harems.

However, physical fighting is not the only means by which males compete with each other. When a coalition of male lions successfully takes over a harem of females, the first thing the males do is to kill all the young lion cubs in the pride. The cubs, of course, were fathered by the previous males, and are no loss to the new owners. There is even a gain to them. While a lioness is lactating for her cub she will not produce another cub; but when her cub is killed (or is weaned) she soon becomes ready to reproduce again. By killing the cubs, the males bring forward the time when they can start reproducing. Infanticide by males which have just taken over a harem is probably common in nature in many species: it has been observed several times in the Hanuman langur (*Presbytis entellus*, a species of primate) and has been anecdotally recorded in many other mammals.

A more subtle, less gory, form of male competition is the microscopic battle fought among the sperms of different males. In many species, if a female mates with two males, the second male by some means or other manages to fertilize many more than half the eggs. In the fruitfly *Drosophila* for example, the second male fertilizes from 83% to 99% of the eggs. Males have evolved other kinds of counter-adaptations to prevent sperm competiton. Some male damselflies stay with their mates after copulation and fight off any other males who come near; only after the female has laid her eggs does the male leave her. Likewise, the acanthocephalan worm *Moniliformis moniliformis* male sticks a 'chastity belt' on a female after mating, probably to prevent other males from copulating with her. The males are also known to 'rape' other males, cementing up the victims' genital openings to render them incapable of copulation. An extraordinary kind of sperm competition has been found in the hemipteran insect *Xylocoris maculipennis* by J. Carayon. Copulation in this species is achieved by injection, the male simply punctures the side of the female with his genitalia and squirts his sperm in. However, a male sometimes copulates with another male. His sperms then migrate to the victim's testes, and when the second male comes to copulate with a female, the first male's sperm will be injected into her.

8.4 Courtship and female choice

The classic example of a courtship sequence was worked out by Tinbergen, in the three-spined stickleback. The male stickleback defends a territory (p. 149), and if a female wanders into it he usually attacks her. The male has a different appearance from the female: the male's belly is red, whereas that of the female is a glossy silver, distended (if she is ready to lay) with eggs. If a red-bellied stickleback stays in the male's territory he continues to attack it. But if a silvery, bulging-bellied stickleback stays he soon recognizes it as a female, and changes from attack to courtship. The first stage of the stickleback courtship is the male 'zigzag' display. The male swims rapidly back and forth many times. The female, if receptive, responds to this by a 'head up' display. This stimulates the male to lead her to his nest and show her the entrance by pointing his snout into it. The female may then enter the nest and lay her eggs (Figure 8.5).

One thing that ethologists would like to know about this courtship is why it takes the form that it does. Why all this swimming back and forth? Why not something else equally apparently arbitrary, such as blowing bubbles, or dropping pebbles? Tinbergen suggested that the zigzag display results from the conflict felt by the male between attacking the female and fleeing from her. It swims forward to attack, and away to flee. Whatever its origins, the zigzag display is necessary if the female is to be attracted to the nest. Experiments have shown that males which perform the zigzag display at a higher rate (more swims back and forth per minute) are more likely to be successful at courtship: a female is more likely to mate with a male whose zigzagging is more energetic. In another experiment, by Semler, females were allowed to choose between mating with a normal red-bellied male or with a male of the

Figure 8.5 Courtship usually consists of a recognizable sequence of male and female activities, ending in mating. Courtship in the three-spined stickleback begins with a 'zigzag' display by the male; if the female responds by showing a silvery egg-filled belly, the male leads her to his nest, and shows her the entrance. The female may then enter the nest, and lay her eggs while the male nuzzles her tail. After the female has laid her eggs and left the nest, the male follows her through and fertilizes the eggs. (After Tinbergen)

ordinary non-breeding stickleback green colour. (In certain lakes in North America, there are male sticklebacks whose bellies do not turn red.) The females usually chose to spawn with the red-bellied males. Female choice, therefore, may be the reason why such elaborate courtship displays are found in males. Males continue to court in the way they do because if a male were to stop producing part of the species' typical display, females would not mate with him. Male behaviour which does not attract females to mate is eliminated by natural selection.

Female choice is thought also to explain the evolution of bizarre, highly exaggerated displays, such as those of peacocks and other pheasants. The suggestion was first made by Darwin; but his suggestion was made more explicit by Sir Ronald Fisher. It is not enough only to point out that female choice might favour a bizarre male trait; if the trait is deleterious, one must also explain why the female preference is not then selected against, for the females will produce sons with the bizarre trait. Fisher imagined the following evolutionary sequence. Initially the males with slightly longer than average tails might have been fitter than average; females who preferred to mate with them would have an advantage because they would produce better sons. Because the preference is favoured in females, it will become commoner. Now the advantage to longer-tailed males will increase, as they are both fitter and preferred by the females of the population. Under this twin pressure, tail length will increase; female preference and male character will reinforce each other, and evolve together in a 'runaway' process. Male tail length could now evolve to so great a size that it decreased the male's viability, for now that a preference has been established among the females, it can balance a disadvantage in the preferred character. Most subtly, Fisher pointed out how this balance of preference and disadvantage could be stable. Any female who did not prefer males with the deleteriously exaggerated trait would indeed produce fitter sons than other females; however, no one would mate with them. Nearly all the females prefer males with ridiculous long tails: therefore there is an advantage in fitting one's sons to the fashion. The mathematics of the equilibrium are still by no means certain, but let us pass that problem by, to consider an experimental test of the theory.

If the theory is correct, the females in a species like peafowl must prefer those males with the most exaggerated tails. If that preference is not present, there would be no balancing force to counteract the advantage of smaller tail size, and the equilibrium should break down. The crucial test is on the preference; if it does not exist, the theory is wrong. Until a few years ago this test had not been performed in any species with a bizarre male form, and the theory remained a speculation, tentatively accepted by most ethologists in the absence of any plausible alternative. But in 1981 Malte Andersson tested for a preference in long-tailed widow birds, and provided the first strong evidence for Darwin's theory. Long-tailed widow birds live in Kenya; the males have, among other sexual modifications, a far longer tail than the female. Do females prefer to mate with males of longer tail-length? Andersson tackled the question directly by experimentally altering the tail lengths of the males. The males fell into four categories: nine had their tails cut off and a longer one stuck on (with quick-acting glue), nine had a shorter one stuck on, and, as controls, nine were left intact, and nine had their own tail cut off and then immediately glued back again. Andersson measured the attractive power of

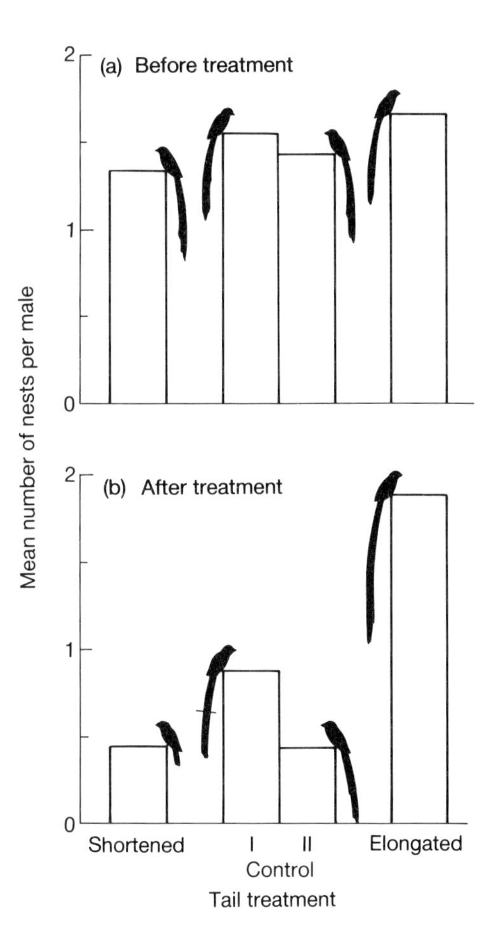

Figure 8.6 Female long-tailed widow birds prefer to mate with longer tailed males. (a) At the top are the mating successes of four groups of males before experimental treatment: they are similar. (b) The males then either had their tails shortened, lengthened, cut off and replaced (control I) or left alone (control II). The males with lengthened tails now enjoyed higher mating successes, the males with shorter tails were less successful. (After Andersson)

the 36 males before and after the treatment; and, to everyone's delight, the number of nests on the territories of the artificially-lengthened males' territories was higher (Figure 8.6). Actually, the experiment is not perfect because, strictly speaking, Andersson only showed that the number of nests on a male's territory was related to tail length; the eggs could have been fertilized by another male, and the female then attracted to the territory of a male with a longer tail. However, in the absence of evidence, this is a pedantic objection. Andersson's experiment strongly supports Darwin's theory.

We only have evidence for one species, but most ethologists who study sex do not hesitate to extend the theory to all bizarre sexual characters. This is not because they are rash. The theory of Darwin and Fisher is the only coherent explanation we possess for such characters. Male competition can explain many structural and behavioural differences between the sexes, but not those that are deleterious to one sex: male competition can favour characters that make males stronger but not merely ornamental characters that would reduce their fighting abilities. It takes careful study to confirm whether a particular sex difference is indeed maintained by male competition or female choice; but the considerable evidence now available suggests that Darwin was essentially correct. Sexual selection is a widespread natural process, and the cause of most behavioural sexual differences.

8.5 Systems of reproduction

The mating system of a species is the way in which the sexes associate for breeding. There are many different kinds of mating systems in the various species of animals. At one extreme, there are species that live in the sea and never move, such as bivalve molluscs, and others that do move, such as starfish, in which the sexes do not really associate for breeding at all. They simply discharge their sperms and eggs into the water, where they fuse. In most species that move, however, the sexes form a definite association for breeding. We shall be concerned here to establish what the main kinds are, and also to ask why they take the form, and occur in the species, that they do.

There are three main categories of breeding system: monogamy, polygyny, and polyandry, which we can take in turn. In monogamy, a single male pairs with a single female. Monogamy is rare in most animal groups, but is common among birds. Over 90% of bird species are monogamous. In a typical monogamous bird like the robin, the male and female stay more or less apart during the nonbreeding season, and only pair up shortly before breeding. They co-operate in preparing the nest, after which the female lays the eggs; both parents take turns in incubating the eggs and, after the eggs have hatched, they both bring food for the young. Similar breeding systems, with biparental care and monogamy, are used by some cichlid fish, and several species of primate. In the odd invertebrate there is monogamy but no ordinary parental care: the limnorid isopods that bore into shipwood live in pairs, but the young look after themselves; likewise, some wood-boring scolytid beetles live in pairs but have no parental care in the ordinary sense of the word — although the young may benefit from the proximity of their parents.

In polygyny, in its purest form, a single male mates with several females.

A minority of males accomplish most of the mating; and many males die bachelors. Polygyny, often combined with a degree of polyandry (as each female may mate with several males), is the commonest breeding system, being found in most arthropods, fish, amphibians, reptiles and mammals. Polygyny in which most females mate with only one male occurs when the females live in 'harems', as, for example in a species of fish, the wrasse *Labroides dimidiatus* (Figure 8.7). Males of this species hold harems of three to six females, and forcibly prevent other males from mating with them. *Labroides dimidiatus* has the additional interesting habit of sex change. When the male of the group dies, the biggest female in the harem quickly changes

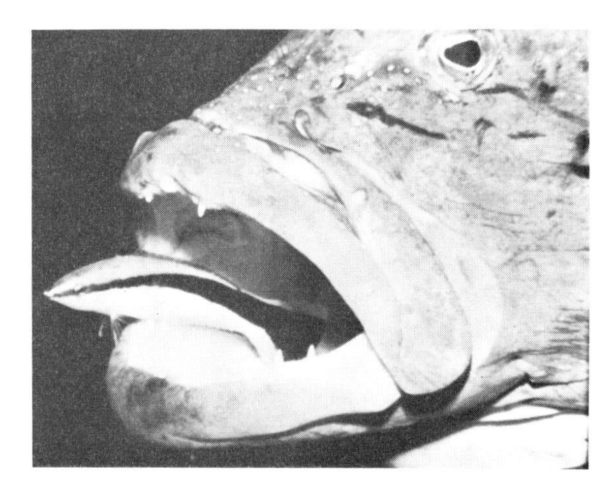

Figure 8.7 The coral reef inhabiting wrasse *Labroides dimidiatus* is a cleaner fish. (a) A wrasse emerging from the mouth of a grouper (*Plectrotromus maculatus*) which it has been cleaning up. The grouper allows the wrasse to enter its mouth, to take out skin parasites and any other edible items. The grouper then allows the wrasse to come out unharmed, and swims away. The grouper loses its parasites, the wrasse gains a meal. The wrasses are territorial, and the same grouper may later return to the same cleaner wrasse. On the wrasse territory, a male defends a harem of females; when the male dies, one of the females changes sex and takes over the harem. (b) Territorial encounter between neighbouring males. (Photos: D. Ross Robertson)

Figure 8.8 Sage grouse mate on leks, in which males aggregate and defend territories from each other and are visited by females. The males are larger than the females and have modified tail feathers (left) and enlarged white oesophageal sacs which they puff up during the 'strut' display (right). (After photos of R. Haven Wiley)

into a male and takes over the group. Another kind of polygyny is often found in species in which the males defend territories that are briefly visited by females for mating, after which the female leaves and the male awaits the next mate. There are no harems in this kind of polygyny; males only associate briefly with females, one at a time; and one female may sample several males. The most extreme forms are those species that have so-called 'lek' mating systems (Figure 8.8). The species of bird called the ruff (*Philomachus pugnax*) is an example. There are relatively few territorial sites, which the strongest male ruffs occupy; these sites, called leks, tend to remain in the same place every year. The leks are visited by the female ruffs (called reeves), and the few males holding territories on the lek do nearly all the mating. After a reeve has mated she departs from the lek to lay her eggs and rears her young by herself.

Polyandry, in its pure form, is the opposite of polygyny, although, as we have seen, they are often mixed. In pure polyandry, each male mates with only one female, but a female (if successful) may mate with several males. It is much rarer than polygyny; but some examples do exist. The kind of bird called the jacana is one. The female jacana defends a harem of males; she lays a clutch of eggs for each of her males, and it is the male who incubates the eggs and rears the young, unaided by the female. The female is bigger than the male, and forcibly prevents other females from mating with, or laying eggs for, any of her males.

Now we have seen the main systems of breeding we can ask again why they take the form they do, and why some species are monogamous, others polygynous, others polyandrous. Relatively little progress has been made in answering these questions. One important association is between the mating system and the mode of parental care. Even the few examples we have seen illustrate how species with biparental care tend to be monogamous, species in which only the female looks after the young tend to be polygynous, and species in which only the males care for the young tend to be polyandrous.

The association, moreover, makes sense. Sexual selection only operates because individuals of one sex contribute more to the production of offspring than the other. Normally the productive sex is female. Males then compete, females choose, and the mating system is polygynous. But if the fundamental variable is altered, so too are the theoretical predictions. If both sexes work to produce offspring, as in a species with biparental care, the selection for males to compete for matings is relaxed, and monogamy may result. If males do the majority of the productive work, as in jacanas, the whole pattern of sexual selection may be reversed. The females evolve to be larger and brighter coloured than the males, and control harems of dull-coloured egg-sitters, competing for their domestic skills against other acquisitive and territorial females.

In one sense, the association of parental care and mating system transfers the question back a stage; for why do some species have biparental, others maternal care? We shall come to that in a moment. But more seriously, it provides only a partial solution to the problem. Most species completely lack parental care, but they are by no means uniform in their mating systems. Many are polygynous, as we should expect, but others, like the wood-boring isopods and beetles, are monogamous. Moreover, the association of parental care and mating systems is imperfect. In many fish, like the three-spined stickleback, the males provide all the parental care, but the mating system is polygynous. Other factors are at work.

Let us finish by returning to the question of why species differ in their sexual division of parental labour. The incidence of biparental care presents no great puzzle. It is presumably used by species in which it takes two adults to provide for the young. But what of species in which, it appears, only one parent is necessary. Why is it the male in some species but the female in others? Well, let us first examine some examples. Pycnogonids are small spider-like creatures, ½ inch to 6 inches long depending on the species, which are not uncommon on seashores all over the world, clinging to sea anemones or the bottom of rocks (Figure 8.9). They have eight pairs of walking legs, with their genital openings situated in the top segment. At mating, in an operation never exactly described, the sexes are said to bring their genital openings into opposition and release their gametes. The male has an extra pair of legs at his front end, called ovigerous legs. He uses these to carry the female's eggs, in a ball, for a few weeks. During that time he may mate with other females, and add the eggs to his collection. I have seen males of the genus *Nymphon* on the

Figure 8.9 This male pycnogonid of the species *Boreonymphon robustum* is covered with his offspring, whom he is carrying. Most species of pycnogonids transport their offspring only during the egg stage; but this species carries its young too. Male pycnogonids have a special pair of ovigerous legs; they can be seen here at the front, ending in pincers. (After D'Arcy Thompson)

shores of Britain carrying as many as six bundles of eggs, each from a different female. Such is one kind of parental care: now consider the more familiar habits of fish. In the common river genus *Cottus*, or sticklebacks, or seashore blennies, the male defends a territory; the female visits him, lays her eggs, which he then fertilizes; the female departs; and the male then looks after the eggs. Maternal care is used by many more groups: spiders, crabs, prawns, wasps, frogs, birds, mammals — no one has ever counted them all. They do, however, share one feature different from the paternally-caring fish, and perhaps different from pycnogonids: they nearly all have internal fertilization.

There is indeed a suggestive correlation between the mode of fertilization and the sex that ends up looking after the young. Paternal care tends to be associated with external fertilization, maternal care with internal fertilization. The association is imperfect. In 1976 I counted the number of families of animals with paternal care that had the two modes of fertilization: the result (now out of date) was 42 with external, to 20 with internal fertilization. No one has done the analogous count for all groups with maternal care. My result does show there are exceptions — the jacanas are one — but the rule seems to have enough generality to invite explanation. One possibility was put forward by Richard Dawkins and Tamsie Carlisle. They supposed that one sex was going to care for the eggs, but that selection on both sexes would make

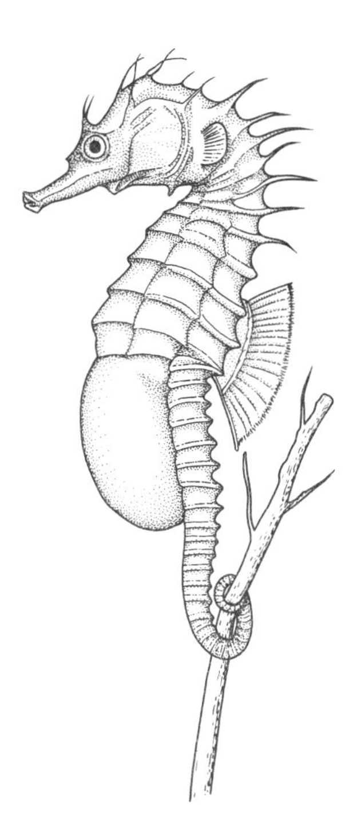

Figure 8.10 Seahorses, such as this pregnant male *Hippocampus antiquorum*, are unique (with some pipefishes) in the animal kingdom in that the males are impregnated with eggs by the female, whose elongated ovipositor introduces the eggs to the male's brood pouch. (After Hesse-Doflein)

them, if possible, desert the eggs and thus force the other sex into doing the work. There would then be an evolutionary race among the sexes to run away first, and leave the other 'holding the baby'. The winner of such a race might well depend on the mode of fertilization. With internal fertilization, in the female, the male could run away before the female could lay her eggs; whereas with external fertilization, if the eggs are released first, the female may be the sex with a head start. The theory is incomplete because, as we have seen, the association it would explain has exceptions. However, it may provide a part of a general theory, which might relate the mode of fertilization, the kind of parental care, and the system of mating, to explain the several general types of reproductive system.

8.6 Summary

The first requirement of a mating animal is to find a partner of the right species; this is accomplished in crickets by acoustic signals. The need to recognize species, however, cannot explain the bizarre development of sexual characters. Darwin proposed his theory of sexual selection to account for sex differences of structure and behaviour, according to which, in most species, males compete for, and females select, mates. The theory can be contained in a coherent modern formulation. Males in nature do compete for females, in diverse ways: obvious physical fighting, and subtle invisible competition among sperms within the female reproductive system. Female choice probably has selected for many of the properties of courtship, such as the 'zigzag' display and red belly of male sticklebacks. Female choice is also the only known valid explanation of extravagant, and probably deleterious, male character such as the peacock's tail. Female choice has only properly been tested for in the case of one such character, the long tail of a species of widow bird, and it was confirmed to be operating. The sexes form different kinds of association for breeding in different species; some are monogamous, others polygynous, others polyandrous. The differences can be partly explained by the habits of parental care; monogamous species often have biparental care, polygynous species maternal care, and polyandrous species paternal care. The difference between species with paternal care and maternal care can in turn be partly explained by whether fertilization is effected externally (which is associated with paternal care) or internally. The explanations, however, are incomplete.

8.7 Further reading

Trivers (1985), R. Dawkins (1976) and Maynard Smith (1978) can be recommended on the grounds of importance, clarity, or both; the collection edited by Bateson (1983) is useful. Darwin (1871) remains an excellent review of the sexual characters of animals, with many thoughtful comments that are still relevant and are of more than merely historical interest.

Bentley & Hoy (1974) describe their work on crickets; Andersson (1982) is the reference for widow birds. The study of mating systems is more question than answer at present, and the questions may not be well put, but Emlen & Oring (1977) are good on birds. Ridley (1978) is my review of paternal care.

9 / Altruism and social life

9.1 The natural selection of altruism

Altruism means the transfer of some benefit from the altruist to the recipient, at a cost to the altruist, and it is characteristic of all of what we think of as the most highly developed forms of social behaviour. This meaning of the term has an important difference from that of ordinary human conversation. When we call a human act 'altruistic' we not only mean that the altruist has given something away, but also that the altruist intended, roughly speaking, to be kind; the word refers to subjective intent as well as the objective transfer of goods. Subjective intentions, however, rightly or wrongly, are almost invariably ignored by ethologists, who concentrate exclusively on observable units of behaviour. In borrowing the word 'altruism', therefore, they have altered its meaning to fit their scientific method. The word 'selfish' has likewise been borrowed and stripped of its subjective connotation: ethologists call behaviour selfish if it confers a benefit on the selfish actor, at some cost to the victim. In ethological terms it is altruistic to give another individual a meal, and selfish to take one away, regardless of what subjective intent may lie behind the deeds. It is a normal part of scientific procedure to take words from ordinary language and modify their meaning; the scientists themselves are not confused by it, although outside commentators sometimes are.

Altruism interests biologists because, at first sight, it seems to contradict the theory of natural selection. Natural selection favours traits that increase the reproductive success of their bearers. Altruistic traits, however, must do the opposite. Indeed they decrease reproductive success *by definition*, because, although I left the measurement of 'benefits' and 'cost' vague above, the only universal currency in which to measure them is the currency of offspring. We may define an altruistic act as one that increases the reproduction of the recipient and decreases that of the altruist; and a selfish act as the opposite, one that increases the reproduction of the selfish type and decreases that of the victim. With those definitions, natural selection ought only to favour selfish behaviour. Altruistic behaviour, however, that fits the definition, does exist. We shall consider some examples in more detail later. Let it suffice, therefore, to mention now the sterile 'worker' castes of social insects, which only work to increase the reproduction of another individual; they do not reproduce themselves. How could natural selection favour sterility?

One answer we must consider and dismiss. It is called the theory of 'group selection'. We have supposed that natural selection favours behaviour that increases the reproduction of individuals; but according to the theory of group selection it does no such thing. It favours, instead, the reproduction of whole groups. If it did, altruism would no longer be puzzling, for altruism probably does increase the reproduction of the group of animals as a whole. The group selectionist imagines some groups made up of altruists, and others of selfish members; the altruistic groups would produce more offspring as a whole; and (it is reasoned) that is why altruism exists. Now no one doubts that altruism increases group productivity. What is controversial is whether natural selection can, for that reason, favour it.

Consider now what is going on within each group. If there is an altruist or two within a selfish group they will leave less offspring than average and the habit will be selected against. Vice versa, selfish individuals within a mainly altruistic group will take advantage of the altruistic deeds performed to them and (by definition) out-reproduce the altruists. Within each kind of group selfish animals are at an advantage and will increase in frequency. Selection on individuals within a group and selection of whole groups are therefore favouring opposite traits: group selection favours altruism and individual selection favours selfishness. Which will win? Most biologists who have considered the question in detail have answered in favour of individual selection. We need only consider the general reason. It is the different rates of the two processes: individual is faster than group selection. Individual selection adjusts the relative frequencies of selfish and altruistic types every generation; group selection can only operate when a group goes extinct. The extinction of groups cannot take place faster than the death of their members, and in nature probably takes place much more rarely. The plodding process of group selection may favour altruistic groups from time to time; but individual selection will quickly re-convert each group to selfishness long before the groups go extinct again. At any one time, most animals' behaviour will be selfish. Models in which group selection wins out require unrealistically high rates of group extinction. (An analogous argument has been used above, for the case of cultural evolution and genetic evolution by individual selection, on page 81.)

If individual selection should prevail in nature, altruism is a stronger paradox than ever. Altruism is exactly what individual selection should act to prevent. We do however have a solution to the paradox, which is mainly due to the work of W. D. Hamilton. It comes from considering how natural selection will operate when interacting individuals are genetically related. Natural selection can only work at all on a character that is heritable and it is easy to see how, in some circumstances, a heritable disposition to provide parental care might be favoured. The caring male stickleback, for instance, fans his eggs in their nest. If the male is taken away, the oxygen concentration in the next declines, many of the eggs catch fungal diseases, and then die; parental care therefore is a benefit to the stickleback's offspring. It is also presumably provided at some cost to the male, for fanning is a vigorous activity. A heritable tendency to provide parental care, present in the male, would be inherited by the very offspring whose survival chances the caring habit was increasing, and thus would increase in frequency. When the heritable tendency first arose, it would be as a single genetic mutation. Because the mate of the individual with the mutant would not have it, by the ordinary pattern of Mendelian inheritance (p. 50) it would be passed on to only half the offspring (Figure 9.1). In order for the mutation to increase in frequency this dilution must be more than made up for by the benefits of parental care. Let us symbolize the increase in survival conferred by care with B (for benefit) and the cost to the male of his exertions C. If the mutant is to increase in frequency, $\frac{1}{2} B$ must exceed C (the $\frac{1}{2}$ being because only half the beneficiaries actually possess the gene). The figure of $\frac{1}{2}$ is what is technically called the 'relatedness', which is the chance that a gene in one (specified) individual is also in another (specified) kind of individual; it is written algebraically as r, and the formula for the condition in which the gene spreads is then $rB > C$ (sometimes called 'Hamilton's rule').

Parental care fits the definition of altruism. The key to an understanding

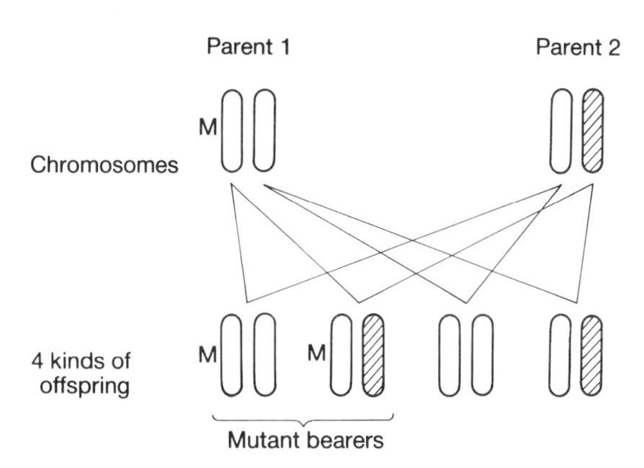

Figure 9.1 When a mutant gene arises it has a half chance of being passed on to an offspring. Offspring can be of four kinds, according to the combination of chromosomes they inherit from their parents, and the four kinds have equal chances of being produced. Two of the four kinds bear the mutant gene. The chance of any particular offspring bearing the mutant gene is therefore 2/4, or 1/2.

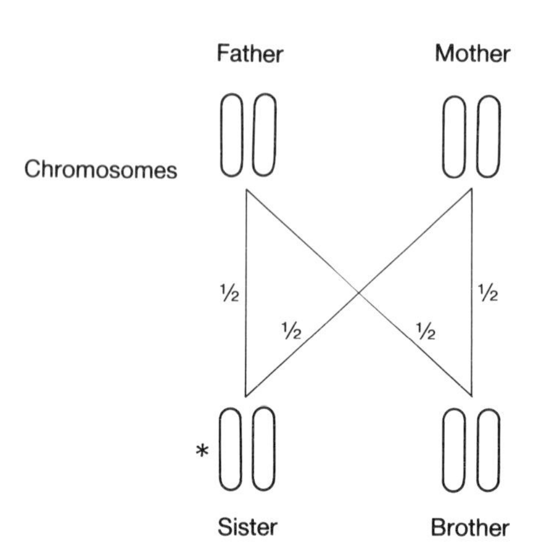

Figure 9.2 Calculation of relatedness from sister to brother under normal diploid Mendelian inheritance. Consider a new mutant gene (*) in a female. What is the chance that it is in her brother? The gene in the female has a 1/2 chance of being in her father, and a 1/2 chance of being in her mother. If it is in her father it in turn has a 1/2 chance of being passed on to her brother, giving a total chance that the gene is shared through her father of 1/4. The chance of sharing the gene through her mother is likewise 1/4. The total chance of sharing the gene is the sum of the maternal and paternal probabilities, 1/2. This is the relatedness of a sister to her brother (see Table 9.1).

of Hamilton's theory is that a genetically equivalent process can operate between other classes of genetic relatives. Let us see what the relatedness (r) is between a full brother and sister (Figure 9.2). Again, we are interested in whether a mutation will spread in the population from an initially rare state. We therefore suppose that it will be in only one copy in any individual. Suppose a female possesses a mutation: what is the chance it is also in her brother? The calculation is this. She has only one copy, which came either from her father (chance ½) or from her mother (chance ½). The chance that a gene in a parent is inherited by its offspring is ½: therefore the total chance of the gene being shared between brother and sister through their father is ½ × ½ and through their mother is also ½ × ½. The total chance of sharing the gene is the sum of these two probabilities, ¼ + ¼ = ½. The relatedness between full siblings is one half. Similar arguments, of increasing complexity, can be made for all classes of relatives; I have listed some of the results in Table 9.1.

Table 9.1 Relatedness (r) among several different classes of relatives.

Classes of relatives	Relatedness
• Parent–offspring	½
• Full siblings	½
• Half siblings	¼
• Grandparent–grandchild	¼
• Uncle/aunt–nephew/niece	¼
• First cousins	⅛
• Second cousins	1/32

Just as a mutation causing parental care will increase in frequency if the care is of sufficient benefit relative to its cost, so can a mutation causing altruism among any class of relatives. Take the case of full siblings. Suppose that a mutation arises, such that mutant individuals direct altruism to their sibs. The mutation, on average, will be present in half the recipients. If it is to spread, the extra reproduction (B) of those individuals must more than make up for the decreased reproduction (C) of the altruist. Again, therefore, $\frac{1}{2}B$ must exceed C. The appropriate figure for relatedness from Table 9.1 can be substituted to give the condition for altruism to be favoured among any class of relatives. Altruism among uncles and their nephews will be favoured provided $B > 4C$. The formula can be easily extended for mixtures of relatives. An act that benefits a group made up of 30% nephews and nieces and 70% offspring will be favoured if $(\frac{3}{10} \times \frac{1}{4}) + (\frac{7}{10} \times \frac{1}{2})\, B > C$. And so on.

Hamilton's theory of altruism makes a clear prediction. Animals should only be altruistic towards genetic relatives, and under quantitatively specifiable predictions. If that prediction is satisfied in real cases of altruistic behaviour, the puzzle of altruism may be removed. If it is not, we shall need another theory. Let us consider two examples where Hamilton's theory seem to apply, and then two others where other processes are at work.

9.2 Helpers at the nest

In many species of birds, the usual breeding unit is a single male and female, who breed and care for their young together. The breeding system of the Florida scrub jay (*Aphelocoma coerulescens*), a bird that breeds in oak scrub in Florida, is rather different. A male and a female form a pair, and the female lays her eggs in the nest, but, when the time comes to look after the offspring, not only the two parents bring food and chase away predators such as snakes; they may be assisted by up to six 'helpers'. A helper is usually a bird that has not yet bred itself, and was produced in an earlier clutch of the adult pair that it is helping; about three-quarters of helpers are the progeny of a previous clutch at the same nest.

The crude prediction of Hamilton's theory is therefore correct: the helpers are genetic relatives. The measurements of Glen Woolfenden enable us to try an approximate test of the precise prediction too. We have seen how to calculate the relatedness r, and that the theory predicts altruism only when $rB > C$ (where B is the average benefit and C the cost of the altruistic act). How can we measure B and C? They actually refer to the change in the lifetime reproductive successes of altruist and recipient, relative to the act not having been performed. The actual B and C therefore are unmeasureable, because they refer to a situation that does not exist (namely, if the act had not been performed). They can however be estimated. A first estimate comes from the breeding success of groups with and without helpers. The choice open to a helper is either to help or breed alone. The cost of helping (C) is the number of offspring it would produce if it had bred alone and the benefit (B) is the number of extra offspring produced by a breeding pair by virtue of having a helper

Woolfenden measured the number of offspring produced by 'experienced' (had bred before) and 'inexperienced' (first time) breeders with and without helpers (Table 9.2). To estimate the benefit and cost for an individual helper we should need to know the breeding success of pairs with one, two, three, etc., helpers rather than for an average of 1.7 or 1.9 helpers as in the table. However, we might approximately estimate the benefit of an individual helper's altruism by dividing the number of extra offspring by the number of helpers. This gives $B = (2.06 - 1.03)/1.7 = 0.61$ offspring if the helpers help an inexperienced pair and $B = 0.31$ for experienced pairs. (Which is probably an underestimate, as there is probably a 'diminishing return' as extra helpers come along; simple division underestimates the benefit of the first 'marginal' helper.) The appropriate figure for cost, C, is simply the breeding success of an unhelped inexperienced pair, which is 1.03 young. If we take $r = ½$ we can readily see that for either estimate of B, $rB < 1.03$ (it is either 0.3 or 0.15) with these estimates, helping at the nest does not satisfy Hamilton's rule.

The test, however, was not a strong one. We have made several dubious assumptions, one of which is clearly false. The helpers of inexperienced pairs cannot be assisting to produce siblings; r is therefore probably less than one-half. Three other assumptions are as follows. One is that the extra offspring produced by helped pairs are due to the helping — they might not be. The helped pairs might have a higher breeding success than the unhelped pairs

Table 9.2 Breeding success of Florida scub jays (from Emlen who used the measurements of Woolfenden).

	Young reared by pair		
	without helper	with helper	**Average number of helpers**
Inexperienced pairs	1.03	2.06	1.7
Experienced pairs	1.62	2.2	1.9

for some other reason: to test whether the extra offspring are due to the helping, the experimental addition of helpers would be needed. However, it seems likely that the helpers do increase the reproductive success of the breeding pair; they have, after all, been seen physically helping, by bringing food to the young. A second assumption is that the survival rate of helped and unhelped pairs is the same. The measurements were made within a season, but the B and C of Hamilton's rule refer to lifetime effects: season differences only measure lifetime effects if survival rates are the same. If helped pairs survive better (or worse) than unhelped pairs, the estimates of B should be modified. In fact Woolfenden recorded a 20% mortality rate per year among unhelped pairs, but only a 13% rate among helped pairs; the extra survival may be an additional benefit of helping. A third assumption is in the estimate of cost. We have assumed that if an individual did not help, it could achieve the breeding success of an inexperienced, unhelped jay; but it probably could not in fact. Woolfenden points out that territories, needed for breeding, are in short supply. Helpers tend to inherit the territory of a pair they have helped, when one of the pair dies. A helper that left its breeding group to try to breed alone would probably therefore produce on average much less than 1.03 offspring; the true value of C is probably much less than our estimate.

Therefore, although Hamilton's rule can in principle be tested quantitatively, in practice there may be many difficulties in estimating the benefits and costs of an altruistic act. The behaviour of Florida scrub jays does look roughly in line with the theory. Kin selection (as the process embodied in Hamilton's rule is called) has probably influenced the evolution of the habit. But the need of the birds to inherit a territory may have been at least as strong an influence.

Florida scrub jays are by no means the only bird species with 'helpers at the nest', or animal species with co-operative breeding; indeed the test is now expanding rapidly enough to put in doubt the generalization that most bird species breed in monogamous pairs. At present, over 150 bird species with helpers at the nest are known. Some mammal species, such as the red fox, the mongoose, the African hunting dog, and the jackal, have been found to have a breeding unit of one pair with a number of helpers (Figure 9.3). Not all the species with helpers at the nest have the same social system as the Florida scrub jay, however. In some, such as the ostrich (*Struthio camelus*) and another kind of bird called the groove-billed ani (*Crotophaga sulcirostris*), several females lay eggs in the same nest; in the ostrich, only one of the

Figure 9.3 'Helpers' in two mammal species. (a) An adult male grey meerkat (*Suricata suricatta*) in the Kalahari, southern Africa, is babysitting. The offspring are of the dominant female of the group and this baby-sitter is probably their brother or half-brother. The social behaviour of this species has been observed by David Macdonald (whose work is not yet published). The adults not only babysit the offspring of the group — forgoing their own foraging time to do so — but also feed the young and seem to teach them to forage; each youngster attaches itself to one adult for instruction. (b) A pair of female red foxes in England. The females are sisters and living in the same group. In general only one female, the most dominant one of the group, breeds at a time, and some non-breeding females help to rear the offspring, by giving them food, grooming and playing with them, and retrieving them if they stray from the den. (Photos: David Macdonald)

females incubates all the eggs. In other species, such as the dwarf mongoose (*Helogale parvula*), many of the helpers are not genetic relatives of the breeding pair who they help. Helping in mongooses, therefore, cannot be explained by Hamilton's theory of kin selection.

9.3 Insect societies

9.3.1 The social life of insects

Ethologists honour four groups of insects — ants, bees, wasps and termites — with the description of 'social' insects (Figure 9.4). The crucial characteristic of these four groups is that, within their nests, they show a reproductive division of labour. Within the nest of a typical ant species there is a single queen who lays nearly all the eggs that are laid in the nest. The rest of the ants are sterile 'workers'. The workers carry out all the duties necessary to keep the colony going except for the laying of eggs.

There are more than 12,000 social insect species, and they show a fascinating range of ways of life. Within the ants, for example, there are 'army' ants with huge colonies of up to 22 million individuals, which bestride the jungle floor eating everything edible in their path. There are 'fungus

Figure 9.4 *Polistes*, such as this *Polistes gallicus* nesting on an agave, live in relatively simple societies and build small nests. (Photo: Heather Angel)

gardening' species which grow fungus on specially prepared rotting leaves and live on the produce of the fungus. Other ant species live by milking honeydew from herds of little insects called aphids. In yet others, workers form living 'honeypots', as they hang upside down from the roof of their nest, their abdomens hugely distended with honey. Australian aborigines dig up the nests, take the ant's head between their fingers, and bite off the honeyed abdomen. The marvels of the social insects are almost endless, but we shall concentrate on some general properties of their social life that are closely related to their altruistic habits.

We must consider ants, bees and wasps separately from termites. Ants, bees and wasps belong to the insect order Hymenoptera; termites make up the order Isoptera. A typical hymenopteran colony is founded by a single queen after her 'nuptial' flight. If the species is an ant, the foundress bites off her wings — she will never fly again. She excavates a small nest, lays her first brood, feeds and rears the larvae. She will never rear larvae again, because this first generation of workers themselves rear the next lot of eggs. It is an important fact that all the workers, in the first and later generations, are females. They are sterile and therefore in a sense sexless, but they contain the genetic make-up of females. Once the colony has reached a certain size the queen lays eggs which are reared as reproductives. The time when reproductives are first produced varies between species. In *Myrmica rubra*, a common garden ant in Europe, it is not until about 9 years after founding, when the colony has grown to about 1000 workers.

9.3.2 Altruism in insect societies

The distinctive property of social insects that cries out for explanation is the sterililty of the workers. Why do these workers slave away to their death only to enhance the reproduction of another individual? An important part of the answer may be provided by Hamilton's theory. For Hymenoptera have an idiosyncratic genetic system, rather different from the usual Mendelian one. They are 'haplodiploid'. The females, like most animals, have two sets of chromosomes (they are 'diploid'); but the males have only one (they are 'haploid'). Male offspring contain no genes from their fathers; they develop from unfertilized eggs (Figure 9.5). Females contain genes from both their mother and their father. Now haplodiploidy, as Hamilton pointed out, will alter the normal pattern of relatedness. The sisters of one family are

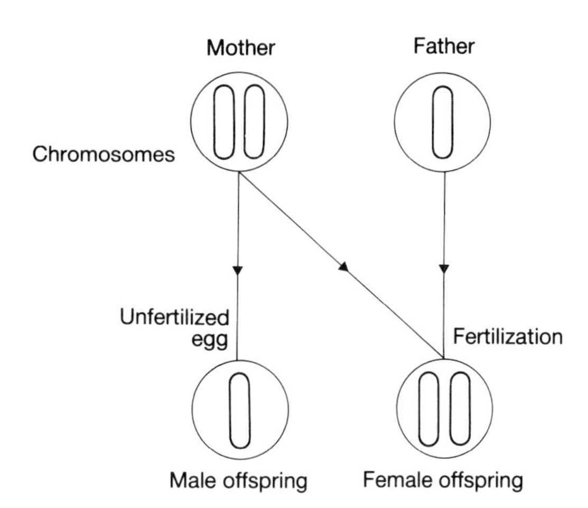

Figure 9.5 Haplodiploid inheritance. Females have a diploid set of genes but males have only one ('haploid') set. Males develop from unfertilized eggs. This kind of inheritance is exceptional, but is found in hymenopteran insects and a few other groups of animals. The relatednesses among individuals differ from those under diploid inheritance: a gene in a male, for instance, has a probability of one of being in his mother (see Table 9.3 and contrast with Table 9.1).

exceptionally closely related because they all share exactly the same set of genes from their father; their father has only one set of genes to give. Many of the values of relatedness are changed under haplodiploidy (Table 9.3). The relatedness between siblings under diploidy was calculated under the premise that if a father contributes a gene to one sibling there is only a chance of one half that he will give it to another. Under haplodiploidy that chance is one, not a half, and the total relatedness between sisters is therefore $(\frac{1}{2} \times 1) + (\frac{1}{2} \times \frac{1}{2}) = \frac{3}{4}$. Because the relatedness of mother to daughter is the same as under diploidy (one half), a female is actually more closely related to her sisters than her daughters, and can, other things being equal, 'breed' more efficiently by making sisters than by reproducing daughters. That may be why female hymenopterans have so often evolved sterility. Males, by

Table 9.3 Relatedness under haplodiploidy. The relatedness is the probability, given a gene is in one kind of individual, that it is in another kind of individual. They can therefore be asymmetrical, as between mother and son.

Relationship			**relatedness** (r)
•	Mother	daughter	½
•	Mother	son	½
•	Father	daughter	1
•	Father	son	0
•	Daughter	mother	½
•	Son	mother	1
•	Brother	sister	½
•	Brother	brother	0
•	Sister	sister	¾
•	Sister	brother	¼

constrast, have not evolved sterility. But then they are not exceptionally closely related to their siblings.

A large controversial literature has grown up around Hamilton's explanation of sterility in hymenopterans. It would be inappropriate to consider it here. No ethologist would doubt that kin selection has had some part in the evolution of hymenopteran social behavour, but this statement leaves plenty of room for disagreement. The same exact theory cannot apply to termites. Termites are diploid, therefore the relatedness of siblings is normally the same as that of parents to offspring. But sterile castes have evolved in termites. As Hamilton's theory predicts, in termites both sexes of offspring become sterile workers; in termites, unlike Hymenoptera, there is no force of kin selection predisposing one sex to evolve sterility. However, although the theory can explain this difference from Hymenoptera, there is no widely accepted explanation of the evolution of sterile castes in termites. Suggestions have been made. The relatedness among siblings can be higher than that of parents to offspring if siblings are produced by highly incestuous matings but the offspring produced by outbreeding. A termite that 'went alone' might outbreed, whereas it could help to rear closely-related, inbred siblings if it stayed on as a helper. However, the idea is theoretical only, and is not well tested. I only mention it to illustrate that the theory of kin selection has been applied to termites as well as Hymenoptera. The theory is more easily applied to Hymenoptera.

9.4 Recognizing relatives

We have not proved beyond doubt that kin selection has caused the evolution of altruism in avian helpers and social insects, but the theory probably contains a large measure of truth in these cases. We might pause, therefore, to ask how animals actually recognize their genetic relatives. For if animals are to act altruistically towards relatives, they must be able to recognize them. How do they do so? No work has been done on helpers at the nest. They could 'recognize' relatives merely as other young birds in the same nest as they were reared in as they would usually be siblings. In practice, recognition may be more sophisticated, but in the absence of facts speculation is unnecessary, because there is no difficulty of principle here. There has, however, been some work on insects which merits mention. Colonies of social insects have been known for many years to possess individual colony odours. Individuals can tell members of their own colony (who are genetic relatives) from members of other colonies (unrelated) by their characteristic smell. The source of colony-specific odours is not completely known; but it probably develops from the diet of the colony, as individuals frequently regurgitate food to one another, and minor differences in the diets of different colonies would give them different characteristic odours. Different colonies perhaps secrete their own pheromones as well, but the importance of diet has been indicated by experiment. Kalmus and Ribbands moved two hives of honeybees from a typical honeybee environment to an isolated moor which had only one species of flower. The level of fighting between the two hives decreased, perhaps because they increasingly came to recognize each other as members of the same hive. Likewise, when parts of a hive were isolated and fed on different diets, the level of fighting within a hive increased.

Each member of the colony has to learn the colony odour. Young ants do not challenge other ants in the colony, regardless of their odour; but older ants challenge any ant that does not have the colonial smell. The learning may even take place during a sensitive phase of imprinting.

What I have said so far would probably have passed until quite recently as a short summary of how social insects recognize nestmates. It was supposed that relatives were recognized by cues of environmental origin, by learning. A rush of recent evidence now suggests that story is incomplete. In several species, individuals have been found to be capable of distinguishing relatives even though they had no chance of learning to recognize them: they seem to possess an innate ability to distinguish relatives. The first important experiment was conducted on sweat bees, *Lasioglossum zephyrum*, by Greenberg in 1979. By various genetic tricks, he bred colonies in the laboratory that had twelve different kinds of relatedness to one another, varying from completely unrelated colonies ($r = 0$) to different colonies produced from inbred lines ($r \simeq 1$). In his experiments he introduced a bee from one colony to the entrance of another colony. At the colony entrance a 'guard' bee generally admits nestmates and challenges foreigners, and Greenberg observed the rate at which guard bees admitted introduced bees bearing differing degrees of relatedness to them (Figure 9.6). His result was strikingly positive: guard bees were more likely to admit bees the more closely related the introduced bee was. The important part of Greenberg's experimental design was that in nearly all cases the guard bee had never had any opportunity at all to learn the introduced bee's smell: in nearly all cases guard and

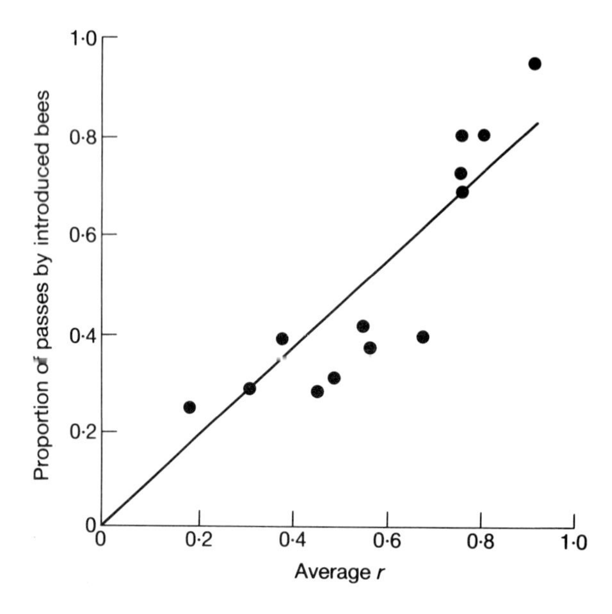

Figure 9.6 Whether a guard bee of *Lasioglossum zephyrum* allows a conspecific bee to enter its nest depends on the genetic relatedness of the two bees: closer relatives are more likely to be admitted. (After Greenberg)

introduced bee had spent all their lives in different colonies. For a minority of cases the bees would have been separated at the larval stage from the same colony, but no learning is thought to take place this early, and, in any case, the positive result still stands even if this minority of experiments is ignored. Sweat bees, it appears, can recognize degrees of relatedness among conspecifics of which they have no experience at all. Since Greenberg's work on sweat bees, moreover, a comparable ability has been demonstrated with varying degrees of certainty in other bees, and other animals, such as ants, mice, quails and tadpoles. In now seems reasonable to guess that genetic abilities to recognize relatives are widespread. Our original understanding must therefore be modified. Animals can recognize their relatives by a mixture of innate and learned information, in ways that probably differ among species and are not, at this early stage of research, well understood.

9.5 Primate societies

9.5.1 The social life of primates

Different species of primates live in different kinds of societies; indeed, the same species may form different kinds of social groups according to the conditions. The Hanuman langur (*Presbytis entellus*) forms both harems, with one male and several females, and multi-male groups, with several males and a larger number of adult females. The reason is uncertain, but may be related to population density, for multi-male groups are commoner when the total population density of an area is lower. Before considering the altruistic behaviour of primates, let us consider briefly some of the different kinds of social groups.

The white-handed gibbon (*Hylobates lar*) lives in the trees of South East Asia in monogamous family groups. The male and female are similar in size and appearance, and live as a pair with 0–4 young. The pair defend a territory of about half a square mile. Within the family there is no dominance or aggression: male and female live together peaceably as equals. Other primates too, such as the spider monkeys (*Ateles*, Figure 9.7a) also live in family groups. But how different are the societies of hamadryas baboons! The hamadryas baboon (*Papio hamadryas*) is a terrestrial species, inhabiting the plains of North East Africa and South West Arabia. The hamadryas baboon society has several levels of organization. The lowest level, the breeding unit, is a harem of one to ten females with a single male. The male treats his females aggressively, often attacking them if they stray away. The male is about twice as large as the female, which influences the dominance relations of the sexes, but the large size of males is probably mainly an adaptation for fighting off enemies, especially rival males. At the next level, several of these harems may walk around together in larger bands while feeding, and these bands may act as a unit to defend a food source from a rival band. The hamadryas baboon, however, does not defend a territory. The bands do confine their wandering to a large area of about 12–15 square miles, but different bands may overlap in their use of this area. It is therefore called a home range, to distinguish it from a defended area, which would be called a territory. At a higher level, several bands may join into larger groups for sleeping. If suitable shelters for sleeping are difficult to find, as many as 700 hamadryas may sleep together.

Whereas in the hamadryas baboon the mating unit is a single harem aggressively controlled by a single male, in the howler monkey (*Alouatta*), several adult males may live peaceably together in a single group (Figure 9.7b). When Ray Carpenter watched howlers on Barro Colorado Island, in Central America, the group sizes were variable but contained an average of 3 males, 8 females, and 7 young. The male howler monkey is about 30% larger than the female, and has an enlarged voice box covered by a beard. The males roar daily, making the loudest animal noise in the American forests, a noise

Figure 9.7 (a) Spider monkeys such as this family of *Ateles belzebuth* living in Panama, live in family groups of a breeding pair and their offspring. (b) Howler monkeys (*Alouatta*), however, live in groups which may contain several adult males and females. Howler monkeys, like spider monkeys, inhabit central American forests. (photos: Fritz Vollrath and Heather Angel)

which carries for over a mile. This howling serves to space out the different groups. Within each group there is little aggression and no obvious dominance hierarchy.

These three species live in entirely different kinds of societies. One is monogamous, another polygamous within single male groups, another polygamous with multi-male groups. One shows dominance and aggression within groups, the other two do not. Two defend territories, the other does not. Other primate species live in yet other kinds of societies. Ethologists would like to be able to explain why different primate species live in different kinds of societies. Why are some species aggressive, others not? Some territorial, others not? These questions cannot yet be answered satisfactorily. Our understanding of the diversity of primate societies is still limited. One trend which can be explained is as follows. There is a tendency in species in which there are many females per male in the group for the males to be bigger than the females: in monogamous species the male is about the same size as the female; in polygamous species the male is larger. This has presumably arisen because sexual selection has favoured larger males in polygamous species, because larger males are more successful in fights over females.

9.5.2 Altruism in primates

Many kinds of altruistic behaviour can be seen in primate groups. There is the feeding, carrying and defence of young, not only by their mothers; co-operative searching for, hunting and exploitation of food; food-sharing; co-operative group defence against enemies; and, most common of all, that favourite pastime of primates, grooming. Most, perhaps all, of these habits should probably be explained by kin selection. We, however, have considered the application of that theory in two examples already, and will therefore use a primate example to illustrate another reason why altruism may evolve: the theory of reciprocal altruism.

The example concerns 'consorting', which is a habit of males, in many species that live in multi-male groups, whereby a male stays close to a female during the receptive phase of her oestrous cycle, and defends her from the advances of other males. It is an adaptation produced by the male competition component of sexual selection. In the olive baboon (*Papio anubis*), Craig Packer observed that two males may occasionally co-operate to fight off a single male: clearly, two will be stronger than one, which makes the advantage of co-operation clear. The advantage, however, is only for the one male that copulates with the female; he has gained a benefit of as much as one extra offspring: the other male has paid a cost of the risk of injury in a fight, but obtains no benefit. He has behaved altruistically. If what we have seen so far were the end of the matter natural selection should eliminate the altruistic habit. But it is not the end of the matter. The next stage arrives when another female in the troop comes into oestrus. The roles of the same two males may now be reversed. Packer saw ten cases in which a male who had previously been 'solicited' into co-operating to defend a female (but did not copulate with her) himself solicited a male into co-operative defence. In nine of the ten cases the solicited male was the individual whom he had previously helped. It looks, therefore, as if males form co-operating pairs to defend females and take turns in the copulating that is the end of the defence. If so, it would be an example of 'reciprocal altruism'. Altruism can evolve under individual selection, without any need for the animals to be related, if the altruist is more than paid back later. The danger of any such reciprocal arrangement is that they will be cheated on. There is a clear short-term advantage to receiving altruism but then not paying it back; that way the cheat gains the benefit but does not pay the cost. This being so, reciprocal altruism is expected mainly to evolve in species that form stable groups, with individual recognition. Then a cheat, after gaining a short-term benefit, can be recognized and excluded from future transactions; the cheating will then not pay. Without the opportunity and mechanism of discriminating against cheats, reciprocal altruism is likely to break down. Because reciprocal altruism requires rather special conditions, it may be rarer than kin-selected altruism. It is, however, a theoretical possibility; and in olive baboons it has been realized in fact.

9.6 Manipulated altruism: the control of behaviour by parasites

Natural selection, we have seen, normally makes animals behave in their own selfish interests. Even when it favours altruistic behaviour, it is only in the interest of some broader form of selfishness. However, conflicts of animals open many opportunities for exploiting other individuals, and animals may not therefore always behave in their own interests. There are probably numerous subtle forms of exploitation within a social group — a possibility we have considered before in relation to animal signals (p. 138) — but the idea of 'manipulation' in animal behaviour is a relatively recent one, and has not been the subject of much research. For clear examples we are forced to go to those unambiguous situations of conflict, parasite–host relations. Here are many examples in which animals do not behave in their own selfish interests, or the interests of their genetic relatives. Hosts under the influence of parasites do show altruistic behaviour, according to our adopted definition, but only because they are in some sense forced to, against their own interest. I should emphasize that I have picked parasite–host examples only because they unambiguously illustrate a process that may be expected to be of wider importance. Many cases of altruistic behaviour in nature may be due to still subtler forms of manipulation than those we are about to consider.

A young cuckoo (*Cuculus canorus*) being fed by its foster parent, such as a reed bunting (*Emberiza schoeniclus*), is a striking example of behavioural manipulation by a parasite. It is not in the reed bunting's interest to feed the cuckoo: natural selection favours reed buntings that rear their own offspring, not cuckoos. To our eyes, it is very strange that a reed bunting cannot tell when it is feeding a cuckoo rather than its own offspring. For the insatiable demands of the cuckoo are still met by its tireless foster parents even after the cuckoo has grown larger than its foster parent. It should be easy, we think, for a reed bunting to distinguish a great, ugly cuckoo from a reed bunting. Yet the parent does not. It continues to pour worms down the cuckoo's all-consuming throat. Something about the continual gaping and begging of the cuckoo compels the reed bunting to continue to provide. The reed bunting is being forced by a parasite to do something against its own best interests. The behaviour of feeding the young, however, is not abnormal, it is just misdirected.

Let us now move on to some stranger examples, in which the parasite actually changes the behaviour of its host. The reasons are to be found in the life-cycles of the parasites. Many parasites grow up in a series of different host species. They may start off in one species, be transferred to an intermediate host, and then to a final host; but there can be a problem in effecting the transfer from an intermediate host to the next one. The normal behaviour of the intermediate host might not take it near the final host. A kind of fluke (flukes are small, flattened, worm-like animals) called *Dicrocoelium dendriticum*, for example, lives in ants as its intermediate host, and sheep as its final host. How is it to get from an ant to a sheep? The trick is to have the ant eaten by a sheep. Ants normally, understandably enough, avoid being eaten by sheep; they stay down in the soil away from grazing sheep. However, an ant harbouring a *Dicrocoelium* changes its behaviour. The infected ant typically contains about fifty *Dicrocoelium* individuals. One of these burrows into the ant's brain and somehow causes the ant to climb a blade of grass and fastens its jaws to the top of the blade; it clings fast until eaten, perhaps by a sheep. The parasite has then reached its goal.

There are many fascinating examples of this kind. Here are two more. Nematomorphs (small worm-shaped animals), live in insects as intermediate hosts, but in water as adults. The parasite has to bring its insect to water, and the nematomorphs do somehow bring this about. In one dramatic record, a bee was flying over a pond when suddenly, as it was about 6 feet above the water, it dived in. As soon as the bee splashed into the water, the worms exploded from the body they had so abused, and swam away, leaving it dead. A final example concerns the parasites of the amphipod *Gammarus lacustris* (Figure 9.8). *Gammarus* is a small shrimp-like animal which lives in freshwater. Normally, *Gammarus* avoid the light, sheltering under pebbles on the bottom. However, when infected by the parasitic worm *Polymorphys paradoxus*, their reaction to light is reversed: they now seek the light and swim just below the surface. The parasite's motive is that the next host after *Gammarus* is a duck that feeds by dabbling at the surface; the *Polymorphus* changes the *Gammarus*'s behaviour to make it be eaten by its final host. The duck gets an easy meal, but it infects itself with a parasite when taking it.

Some biologists and psychologists would not include the manipulated altruistic behaviour in these parasitological examples as examples of real 'altruism'. They would prefer to count only kin selected and reciprocal altruism as real altruism, and exclude from the term behaviour carried out in the genetic interests of others. The process of natural selection does differ between the cases of kin selection, reciprocal altruism, and manipulation, and this might seem to provide a reason to separate them. But, following most authorities, I have defined altruism not by the process of natural selection that favours it, but in terms of its reproductive consequences. An altruistic act is one that increases the number of offspring left by the recipient, and decreases the number left by the altruist. Manipulated behaviour fits the definition of altruism. The consequentialist, non-intentional definition of altruism is far different from the normal everyday usage in which we think of an altruist as intending to be kind. And perhaps some are tempted to include kin selection and reciprocal altruism as true altruism because the

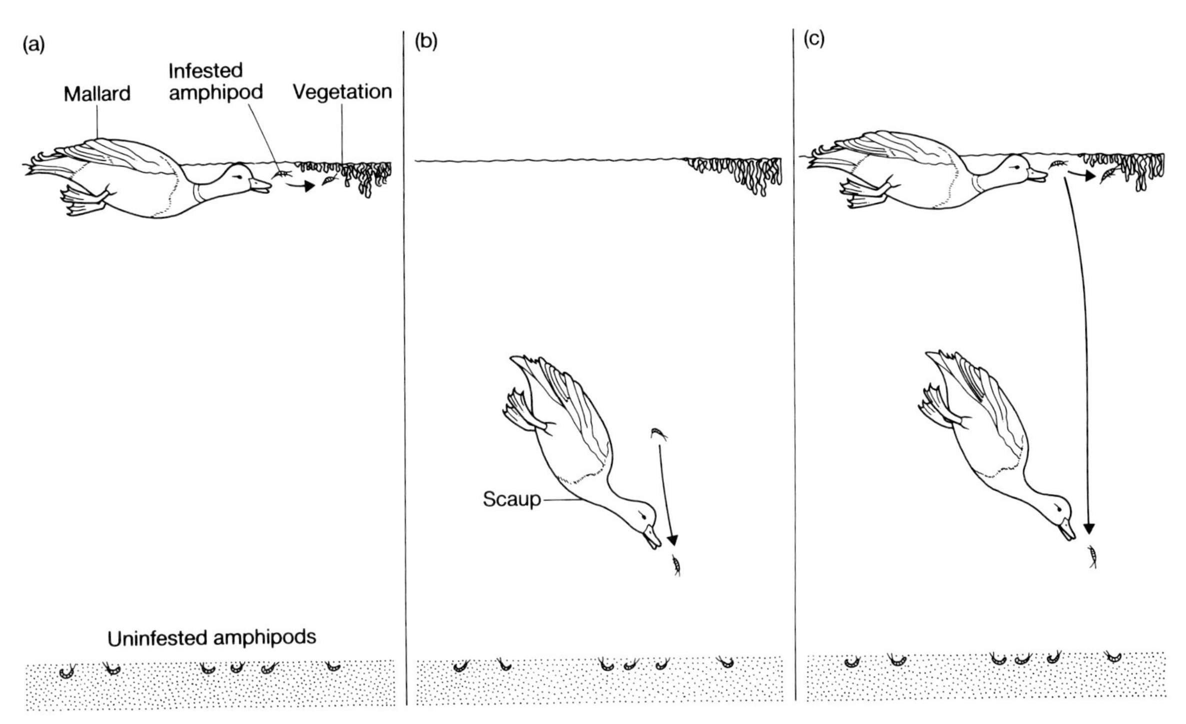

Figure 9.8 Three different parasitic acanthocephalan worms have different effects on the behaviour of the amphipod *Gammarus*. Unparasitized *Gammarus* live in the bottom mud. But if parasitized by *Polymorphus paradoxus* (a), they swim towards the light and come to the surface where they may be eaten by dabbling ducks such as mallards (which are in turn parasitized by the *Polymorphus*). If parasitized by *Polymorphus marilis* (b) the *Gammarus* do not swim to the surface, but do come out of the mud; they are fed on by diving ducks such as scaups. (c) *Gammarus* parasitized by *Corynosoma constrictum* swim to the surface but dive when disturbed; they are fed on by both dabbling and diving ducks. (After Bethel and Holmes)

subjective intention (in the normal meaning of the word) can, by loose and wholly erroneous reasoning, be read into the genetic 'interest' of an indivdual. In kin selection and reciprocal altruism the individual does behave in its *genetic* self-interest (although not its own self- interest). However, it is not part of the definition of altruism that an animal should be behaving in its own genetic self-interest. The word was defined in terms of the effect on the number of offspring produced, regardless of whose genes those offspring carry.

The task of anyone who wishes to exclude manipulated behaviour from discussions of altruism is to redefine the term. Of course, it could be re-defined; but even if it is, the manipulatory examples I have discussed will take on a new use, in illustrating that there is something unsatisfactory in the established meaning of altruism. The examples have an interest of their own, regardless of this matter of definition; but while altruism is defined as any act that increases the reproduction of another and decreases the reproduction of the altruist — and I think this is a satisfactory biological definition for most purposes — then manipulated altruism is altruism. Animals are not always in control of their own behaviour. Sometimes, unable to help themselves, they behave in the interests of other animals.

9.7 Summary

Natural selection should generally favour selfish behaviour, which makes altruistic behaviour a paradox for Darwin's theory. It has been suggested that natural selection can act to the advantage of the group, rather than the individual; but this requires unrealistically extreme circumstances. Natural selection may favour altruism under two known conditions, when it is directed towards relatives and when it is paid back later. The former is probably commoner in nature. It at least contributes to the altruism of 'helpers at the nest' in Florida scrub jays, a species for which we have sufficiently detailed measurements to attempt a quantitative test of the theory. It probably also accounts, at least in part, for the extreme altruism of sterile 'workers' in social hymenopteran insects; Hymenoptera have a peculiar genetic system which predisposes females to evolve the sterile worker habit. If altruism is to be concentrated on relatives, relatives must be recognized. There is much evidence from social insects that nestmates (which are relatives) are recognized by a learned colony odour, the source of which is the characteristic diet of the colony. Different colonies fed on the same diet behave more altruistically towards each other. But an unlearned genetic component to kin recognition has also been shown to exist. Sweat bees behave differently to different classes of relatives, none of whom they have ever experienced before. Many of the kinds of altruistic behaviour seen in primates are probably due to kin selection, but this group also illustrates the theory of reciprocal altruism. Olive baboon males co-operate in pairs, the members of which pay each other back in turn as occasions arise. Altruism can also evolve when it is not in the interests of the altruistic animal. Parasites, such as cuckoos, may manipulate their hosts into behaving in the interest of the parasite but at a cost to the host. This kind of manipulation may, in subtle forms, be widespread in the natural world.

9.8 Further reading

R. Dawkins (1976), Krebs and Davies (1981) and Trivers (1985) introduce the subject of altruism; see especially Dawkins, and Williams (1966) for critiques of group selection. Grafen (1985) is more advanced. The fundamental papers of Hamilton (1964, 1972, 1975) are not always easy reading for the beginner; but they are well worth the effort, not least because the secondary literature contains so many regions of quicksand that can let the unwary reader down. R. Dawkins (1981) and Grafen (1982, 1984) have explained some frequent misunderstandings. My treatment of helpers at the nest mainly draws on Grafen (1984); on the natural history see Emlen (1984). Holmes and Sherman (1983) review the means by which kin recognize one another; and Wilson (1975), among many other things, Chalmers (1979) and Hinde (1974, 1983), for instance, review the social behaviour of primates. Packer (1977) is the reference for olive baboons. My final section is derived from R. Dawkins (1982).

Epilogue

The study of animal behaviour is a relatively young science, dating back to only the first three decades of this century, when biologists in Germany, Holland, Great Britain, and the United States independently started watching animal behaviour closely in nature, aiming to find out why animals behave in the way they do. A huge mass of facts and theories about animal behaviour have accumulated since then, which fall (as we have seen) into four main categories. I am going to finish with some invidious generalizations about the success and interest of those four main currents of research. They are generalizations and they do not cover every case! Broadly speaking, the most successful and enduring work of the 1920s and 1930s was on questions of mechanism: studies of how animals recognize features of their environments, what stimuli release what behavioural responses. The study of development erupted in the instinct controversy of the fities; and since the mid-sixties the study of the survival value of behaviour, by means of the theory of natural selection, has increasingly become the main current of research.

Besides the many successful investigations of behaviour mechanisms, there remain unanswered questions such as that of bird navigation, which we considered in Chapter 4. The study of mechanisms has probably been more successful at the behavioural level, of discovering sensory systems, pattern recognition and sign stimuli, than in reducing the control of behaviour to its neuronal and motivational control within the animal itself, although neuroethology has many discoveries to its credit, as we have seen. The other flourishing and successful area of inquiry has been of survival value, where (as we have seen) several paradoxes in the evolution of social behaviour have recently been solved, at least in principle, and new theoretical concepts have been invented for the study of actual cases. I do not think the study of development has been so successful. Positive discoveries have been made (such as those about imprinting) but the subject has been dominated by the nature/nurture controversy, which, as it has been settled, has left a relatively small residue of positive principle. Moreover, the subject has often been isolated from the natural environment of the species, which has provided the setting of so much successful ethology. With the final question, evolutionary history, the difficulty has been more one of lack of interest than of success, although there are some instances — such as the evolutionary origin of signals — where evolutionary analysis has yielded classic results.

Times can change quickly. Ethology has never been isolated from the rest of biology, and within biology, embryology and phylogenetic analysis are rapidly-expanding fields, which may yet influence the ethological special case. This year's flourishing body of research can quickly ossify into standard formulae, and a now unprincipled collection of observations can suddenly all fall into place. I have emphasized our positive knowledge of animal behaviour, but the health of a science is determined less by its past, or even present, achievement, than by its willingness to tackle unanswered and novel questions. The repertory of facts and principles are of most scientific value as a springboard to propel us into less well known territory.

References

General Works

Alcock, J. (1984) *Animal Behavior*, 3rd edn, Sinauer, Sunderland, Massachusetts.

Hinde, R.A. (1982) *Ethology*, Oxford University Press, Oxford and Fontana, London.

McFarland, D. (ed.) (1981) *The Oxford Companion to Animal Behaviour*. Oxford University Press, Oxford.

Further Reading

Andersson, M. (1982) Female choice for extreme tail length in widow bird. *Nature*, **299**, 818–819.

Baker, R.R. (1982) *Migration: Paths through Space and Time*, Hodder & Stoughton, London.

Baker, R.R. (1984) *Bird Navigation: The Solution of a Mystery?*, Hodder & Stoughton, London.

Barlow, G.W. (1977) Modal action patterns. *In*: T.A. Sebeok (ed.), *How Animals Communicate*, Indiana University Press, Bloomington, Indiana, p. 98–134.

Bateson, P.P.G. (ed.) (1983) *Mate Choice*, Cambridge University Press, Cambridge.

Bell, R.H.V. (1971) A grazing ecosystem in the Serengeti. *Scientific American*, **225** (1), 86–93.

Bentley, D. and Hoy, R.R. (1974) The neurobiology of cricket song. *Scientific American*, **231** (2), 34–44.

Bentley, D. and Konishi, M. (1978) Neural control of behaviour. *Annual Review of Neuroscience*, **1**, 35–60.

Bonner, J.T. (1980) *The Evolution of Animal Culture*, Princeton University Press, Princeton.

Brower, J.V.Z. (1958) Experimental studies of mimicry in some North American butterflies. Part I. The monarch, *Danaus plexippus*, and viceroy, *Limenitis archippus archippus*. *Evolution*, **12**, 32–47.

Camhi, J.M. (1984) *Neuroethology: Nerve cells and the natural behavior of animals*, Sinauer, Sunderland, Massachusetts.

Chalmers, N. (1979) *Social Behaviour in Primates*, Edward Arnold, London.

Clutton-Brock, T.H. and Albon, S.D. (1979) The roaring of red deer and the evolution of honest advertisement. *Behaviour*, **69**, 145–170.

Clutton-Brock, T.H., Albon, S.D., Gibson, R.M. and Guinness, F.E. (1979) The logical stag: adaptive aspects of fighting in red deer (*Cervus elephas* L.). *Animal Behaviour*, **27**, 211–225.

Clutton-Brock, T.H., Guinness, F.E. and Albon, S.D. (1982) *Red Deer: Behavior and Ecology of Two Sexes*, University of Chicago Press, Chicago, and Edinburgh University Press, Edinburgh.

Colinvaux, P. (1978) *Why Big Fierce Animals are Rare*, Princeton University Press, Princeton.

Cullen, J.M. (1972) Some principles of animal communication. *In*: R.A. Hinde (ed.), *Non-verbal Communication*, Cambridge University Press, Cambridge, p. 101–122.

Darwin, C. (1859) *On the Origin of Species*. John Murray, London. (And many reprints.)

Darwin, C. (1871) *The Descent of Man, and Selection in Relation to Sex*, John Murray, London. (Reprinted 1981 by Princeton University Press, Princeton).

Darwin, C. (1873) *The Expression of the Emotions in Man and Animals*, John Murray, London.

Davies, N.B. and Houston, A.I. (1984) Territory economics. *In*: J.R. Krebs and N.B. Davies (eds), *Behavioural Ecology*, 2nd edn., Blackwell Scientific Publications, Oxford, p. 148–169.

Dawkins, M. (1971) Perceptual changes in chicks: another look at the 'search image' concept. *Animal Behaviour*, **19**, 566–574.

Dawkins, M.S. (1983) The organisation of motor patterns. *In*: T.R. Halliday and P.J.B. Slater, *Animal Behaviour. 1. Causes and Effects*, Blackwell Scientific Publications, Oxford, p. 75–99.

Dawkins, R. (1976) *The Selfish Gene*, Oxford University Press, Oxford.

Dawkins, R. (1981) Twelve misunderstandings of kin selection. *Zeitschrift für Tierpsychologie*, **51**, 184–200.

Dawkins, R. (1982) *The Extended Phenotype*, W.H. Freeman, Oxford.

Dickinson, A. (1980) *Contemporary Animal Learning Theory*, Cambridge University Press, Cambridge.

Edmunds, M. (1974) *Defence in Animals*, Longman, Harlow.

Ehrman, L. and Parsons, P.A. (1982) *Behavior Genetics and Evolution*, McGraw-Hill, New York.

Emlen, S.T. (1984) Cooperative breeding in birds and mammals. *In*: J.R. Krebs and N.B. Davies (eds). *Behavioural Ecology*, 2nd edn,

Blackwell Scientific Publications, Oxford, p. 305–339.

Emlen, S.T. and Oring, L.W. (1977) Ecology, sexual selection, and the evolution of mating systems. *Science*, **197**, 215–223.

Ewert, J.P. (1974) The neural basis of visually guided behavior. *Scientific American*, **230** 3), 34–42.

Fisher, J. and Hinde, R.A. (1949) The opening of milk bottles by birds. *British Birds*, **42**, 347–357.

Frisch, K. von (1967) *The Dance Language and Orientation of Bees*, Harvard University Press, Cambridge, Massachusetts.

Garcia, J.F., McGowan, B.K. and Green, K.F. (1972) Biological constraints on conditioning. *In*: A.H. Black and W.F. Prokasy (eds) *Classical Conditioning II. Current Research and Theory*, Appleton-Century-Crofts, New York, p. 3–27.

Gould, J.L. (1976) The dance language controversy. *Quarterly Review of Biology*, **51**, 211–241.

Gould, J.L. (1982) The map sense of pigeons. *Nature*, **296**, 205–211.

Grafen, A. (1982) How not to measure inclusive fitness. *Nature*, **298**, 425–426.

Grafen, A. (1984) Natural selection, kin selection, and group selection. *In*: J.R. Krebs and N.B. Davies (eds) *Behavioural Ecology*, 2nd edn, Blackwell Scientific Publications, Oxford, p. 62–84.

Grafen, A. (1985) A geometric view of relatedness. *Oxford Surveys in Evolutionary Biology*, 2, 28–89.

Hailman, J.P. (1969) How an instinct is learned. *Scientific American*, **221** (6), 98–106.

Halliday, T.R. (1983) Motivation. *In*: T.R. Halliday and P.J.B. Slater (eds), *Animal Behaviour*. 1. *Causes and Effects*. Blackwell Scientific Publications, Oxford, p. 100–133.

Halliday, T.R. and Slater, P.J.B. (eds) (1983) *Animal Behaviour* (3 vols), Blackwell Scientific Publications, Oxford.

Hamilton, W.D. (1964) The genetical evolution of social behaviour I, II. *Journal of Theoretical Biology*, **7**, 1–52.

Hamilton, W.D. (1972) Altruism and related phenomena, mainly in social insects. *Annual Review of Ecology and Systematics*, **3**, 193–232.

Hamilton, W.D. (1975) Innate social aptitudes in man: an approach from evolutionary genetics. *In*: R. Fox (ed.) *Biosocial Anthropology*, John Wiley, New York, p. 133–155.

Harden Jones, F.R. (1968) *Fish Migration*, Edward Arnold, London.

Hinde, R.A. (1974) *Biological Bases of Human Social Behaviour*, McGraw-Hill, New York.

Hinde, R.A. (ed.) (1983) *Primate Social Relationships*, Blackwell Scientific Publications, Oxford.

Holmes, W.G. and Sherman, P.W. (1983) Kin recognition in animals. *American Scientist*, **71**, 46–55.

Kandel, E.R. (1979) *Behavioral Biology of Aplysia*, W.H. Freeman, San Francisco.

Krebs, J.R. (1978) Optimal foraging: decision rules for predators. *In*: J.R. Krebs and N.B. Davies (eds), *Behavioural Ecology*, Blackwell Scientific Publications, Oxford, p. 23–63.

Krebs, J.R. and Davies, N.B. (1981) *An Introduction to Behavioural Ecology*. Blackwell Scientific Publications, Oxford.

Krebs, J.R. and Dawkins, R. (1984) Animal signals: mind-reading and manipulation. *In*: J.R. Krebs and N.B. Davies (eds), *Behavioural Ecology*, 2nd edn, Blackwell Scientific Publications, Oxford, p. 380–402.

Krebs, J.R. and Kroodsma, D.E. (1980) Repertoires and geographic variation in bird song. *In*: J.S. Rosenblatt, R.A. Hinde, C. Beer, and M-C. Busnel (eds), *Advances in the Study of Behavior*, **11**, 143–177.

Krebs, J.R., Stephens, D.W. and Sutherland, W.J. (1983) Perspectives in optimal foraging. *In*: A.H. Brush and G.A. Clark (eds), *Perspectives in Ornithology*, Cambridge University Press, Cambridge, p. 165–216.

Kroodsma, D.E. and Miller, E.H. (eds) (1982) *Acoustic Communication in Birds*, 2 vols, Academic Press, New York.

Lehrman, D.S. (1964) The reproductive behavior of ring doves. *Scientific American*, **211** (6), 48–54.

Lorenz, K. (1958) The evolution of behavior. *Scientific American*, **199** (6), 67–78.

Lorenz, K. (1965) *Evolution and Modification of Behavior*, University of Chicago Press, Chicago.

Lorenz, K. (1966) *On Aggression*, Methuen, London.

McCleery, R.H. (1983). Interactions between activities. *In*: T.R. Halliday and P.J.B. Slater (eds), *Animal Behaviour* 1. *Causes and Effects*. Blackwell Scientific Publications, Oxford, p. 134–167.

Majerus, M.E.N., O'Donald, P. and Weir, J. (1982) Female mating preference is genetic. *Nature*, **300**, 521–522.

Marler, P. (1959) Developments in the study of animal communication. *In*: P.R. Bell (ed.) *Darwin's Biological Work: Some Aspects Reconsidered*, Cambridge University Press, Cambridge, p. 150–206.

Marler, P. and Terrace, H.S. (eds) (1984) *The Biology of Learning*, Dahlem Konferenzen. Springer-Verlag, Berlin.

Maynard Smith, J. (1975) *The Theory of Evolution*, 3rd edn, Penguin Books, Harmondsworth.

Maynard Smith, J. (1978) *The Evolution of Sex*, Cambridge University Press, Cambridge.

Maynard Smith, J. (1982) *Evolution and the Theory of Games*, Cambridge University Press.

Packer, C. (1977) Reciprocal altruism in *Papio anubis. Nature*, **265**, 441–443.

Pyke, G.H., Pulliam, R.H. and Charnov, E.L. (1977) Optimal foraging: a selective review of theory and tests. *Quarterly Review of Biology*, **52**, 137–154.

Ridley, M. (1978) Paternal care. *Animal Behaviour*, **26**, 904–932.

Ridley, M. (1985) *The Problems of Evolution*, Oxford University Press, Oxford.

Roeder, K.D. (1965) Moths and ultrasound. *Scientific American*, **212** (4), 94–102.

Sales, G. and Pye, D. (1974) *Ultrasonic Communication by Animals*, Chapman & Hall, London.

Schaller, G.B. (1972) *The Serengeti Lion*, University of Chicago Press, Chicago.

Schmidt, R.F. (ed.) (1978) *Fundamentals of Sensory Physiology*, Springer Verlag, Berlin.

Schmidt-Koenig, K. and Keeton, W.T. (eds) (1978) *Animal Migration, Navigation, and Homing*, Springer Verlag, Berlin.

Schneider, D. (1974) The sex-attractant receptor of moths. *Scientific American*, **231** (1), 28–35.

Sherry, D.F. and Galef, B.G. (1984) Cultural transmission without imitation: milk bottle opening by birds. *Animal Behaviour*, **32**, 937–938.

Slater, P.J.B. (1978) *Sex Hormones and Behaviour*, Edward Arnold, London.

Smith, J.N.M. (1974a,b) The food searching behaviour of two European thrushes. I, II. *Behaviour*, **48**, 276–302 and **49**, 1–61.

Stabell, O.B. (1984) Homing and olfaction in salmonids, *Biological Reviews*, **59**, 333–388.

Staddon, J.E.R. (1983) *Adaptive Behaviour and Learning*, Cambridge University Press, Cambridge.

Tinbergen, N. (1951) *The Study of Instinct*, Clarendon Press, Oxford.

Tinbergen, N. (1953) *Social Behaviour in Animals*, Methuen, London.

Tinbergen, N. (1958) *Curious Naturalists*, Country Life, London. (2nd edn, 1974, Penguin Books, Harmondsworth.)

Tinbergen, N. (1963) On aims and methods of ethology. *Zeitschrift für Tierpsychologie*, **20**, 410–433.

Trivers, R.L. (1985) *Social Evolution*, Benjamin/Cummings, Menlo Park, California.

Turner, J.R.G. (1977) Butterfly minicry: the genetical evolution of an adaptation. *Evolutionary Bilogy*, **10**, 163–206.

Vines, G. (1981) Wolves in dogs' clothing. *New Scientist*, **91**, 640–652.

Williams, G.C. (1966) *Adaptation and Natural Selection*, Princeton University Press, Princeton.

Wilson, E.O. (1971) *The Insect Societies*, Harvard University Press, Cambridge, Massachusetts.

Wilson, E.O. (1975) *Sociobiology*, Harvard University Press, Cambridge, Massachusetts.

Index